NURSES ILLUSTRATED PHYSIOLOGY

BY

ANN B. McNAUGHT

M.B., Ch.B., Ph.D.

Lecturer in Physiology,
The University of Glasgow.

AND

ROBIN CALLANDER

F.F.Ph., M.M.A.A.

Supervisor of Unit for Medical
Illustration and Photography,
The University of Glasgow.

SECOND EDITION

E. & S. LIVINGSTONE

EDINBURGH & LONDON

1971

AS

ALWAYS

To my husband, James A. Gilmour

Ann B. McNaught

To my wife, Elizabeth

Robin Callander

SBN 0 443 00708 X

First Reprint	1965
Second Reprint	1966
Third Reprint	1968
Second Edition	1971

Printed in Scotland

PREFACE

Henry Ford once said "Nothing is particularly hard if you divide it into small jobs". For the study of Physiology that is what we have tried to do here.

Each page deals with a "small" aspect of the subject and—as with our larger book "Illustrated Physiology"—is complete in itself. In general a lead-picture for each chapter indicates the relative position in the body of the system concerned. On subsequent pages the anatomy of major organs is presented in 3-dimensional form. Microscopic features are shown by inset with approximate magnifications indicated. Associated functions of the various parts are summarized alongside. Thereafter line diagrams are used, where applicable, to show how these functions are carried out. Footnotes indicate the method of regulation and control. Cross references have been inserted to pages where limitations of space make it difficult to incorporate those functions of an organ not directly involved in the work of the system under review.

The book is not intended to replace standard works on the subject much less to supplant lecture notes. Rather it is hoped that it may be thought suitable to augment the latter and perhaps save both Tutor and student nurse the sometimes laborious task of copying blackboard diagrams. Its treatment has been found particularly useful for purposes of revision and review especially for those whose available study time tends to come in small rations.

We are indebted as before to a number of colleagues and sources acknowledged where possible in the text; and to Mr W. G. Henderson and Mr A. D. Lewis of E. & S. Livingstone, for unfailing courtesy and encouragement.

ANN B. McNAUGHT
ROBIN CALLANDER

Glasgow.
June, 1971

CONTENTS

1. The TISSUES (1-14)

The Amoeba 1	Differentiation of Animal Cells 5	Nervous Tissues 12,13
The Phenomena of Life . . . 2	Epithelia 6,7	The Body Systems 14
The Cell 3	Connective Tissues . . . 8-10	
Cell Division 4	Muscular Tissues 11	

2. NUTRITION, METABOLISM and DIGESTION (15-46)

Basic Constituents of Protoplasm 15	Digestive System 27	Small Intestine 37
Source of Energy: Photosynthesis 16	Digestion in Mouth 28	Movements of Small Intestine . 38
Carbon Cycle 17	Control of Salivary Secretion . 29	Absorption in Small Intestine . 39
Nitrogen Cycle 18	Oesophagus 30	Large Intestine 40
Nutrition 19	Swallowing 31	Movements of Large Intestine . 41
Energy-Giving Foods . . . 20	Stomach 32	Nervous Control of Gut Move. . 42
Body-Building Foods . . . 21	Gastric Juice 33	Transport and Utilization of
Protective Foods 22,23	Movements of Stomach . . . 34	Foods 43
Energy Requirements . . . 24,25	Pancreas 35	Heat Balance 44
Balanced Diet 26	Liver (and Gall Bladder) . . 36	Maintenance of Body Temp. 45,46

3. TRANSPORT SYSTEM (47-67)

Circulation (general plan) . . 47	Measurement of Arterial B.P. . 55	Factors required for normal
Heart 48,49	Elastic Arteries 56	Haemopoiesis .62
Cardiac Cycle 50	Capillaries: Exch. of H₂O & Solutes . 57	Blood Groups 63
Heart Sounds 51	Veins: Venous Return . . . 58	Rhesus Factor 64
Origin,Conduction ofHeart Beat . 52	Water Balance 59	Lymphatic System 65
Blood Vessels 53	Blood 60	Spleen 66
Blood Pressure 54	Blood Coagulation 61	Cerebrospinal Fluid 67

4. RESPIRATORY SYSTEM (68-77)

Air Conducting Passages . . 69	Capacity of Lungs 73	Nervous Control of Respiratory
Lungs: Respiratory Surfaces . 70	Composition of Respired Air . 74	Movements. 76
Thorax 71	Interchange of Respiratory	Voluntary and Reflex Factors in
Mechanism of Breathing . . 72	Gases. 75	the Regulation of Respiration. 77

5. EXCRETORY SYSTEM (78-86)

Kidney 79	Regulation of Water Balance . 82,83	Storage and Expulsion of Urine. . 86
Formation of Urine: Filtration . 80	Urine 84	
Formation of Urine: Concentration. 81	Urinary Bladder and Ureters . 85	

6. ENDOCRINE SYSTEM (87-100)

Thyroid 88,89	Suprarenal Medulla 94	Posterior Pituitary 98
Parathyroids 90,91	Anterior Pituitary 95	Pancreas: Islets of Langerhans . 99
Suprarenal Glands 92	Underactivity of Ant.Pituitary . 96	Thymus 100
Suprarenal Cortex 93	Overactivity of Pituitary Cells . 97	

7. REPRODUCTIVE SYSTEM (101-115)

Male Reproductive System . . 101	Ovary in Pregnancy 107	Mammary Glands 112
Testis 102	Adult Pelvic Sex Organs in . .	Relationship between Ant. Pituitary
Male Secondary Sex Organs . 103	Ordinary Female Cycle 108	Ovarian & Endometrial Cycles .113
Puberty in Male 104	Uterine Tubes and Uterus in	Puberty in Female 114
Female Reproductive System. . 105	Cycle ending in Pregnancy. 109	Menopause 115
Ovary in Ordinary Adult Cycle . 106	Uterus 110,111	

8. NERVOUS and LOCOMOTOR SYSTEMS (116-150)

Cerebrum 117	Taste 128	Special Proprioceptors . . . 139
Brain: Horizontal & Coronal Sectn.118	Eye 129	General Proprioceptors . . . 140
Brain: Vertical Section . . . 119	Protection of / Muscles of Eye . 130	Cutaneous Sensation . . . 141
Cranial Nerves 120	Action of Lens 131	Sensory Pathways and Cortex . 142
Spinal Cord 121	Retina 132	Motor Paths. to Trunk and Limbs . 143
Synapse 122	Mechanism of Vision . . 133,134	Control of Muscle Movements . 144
Reflex Action 123,125	Visual Pathways to the Brain. . 135	Locomotor System 145
'Edifice' of the C.N.S. . . . 124	Stereoscopic Vision 136	Skeletal Muscles. 146,147
Sense Organs 126	Ear 137	Muscular Movements . . . 148
Smell 127	Mechanism of Hearing . . . 138	Autonomic Nervous System .149,150

CHAPTER 1.

Introduction:–   The TISSUES

The AMOEBA

All living things are made of PROTOPLASM. Protoplasm exists in MICROSCOPIC UNITS called CELLS. The SIMPLEST living creatures consist of ONE CELL. The AMOEBA (which lives in pond water) exemplifies the BASIC STRUCTURE of all animal cells and shows the PHENOMENA which distinguish living from non-living things.

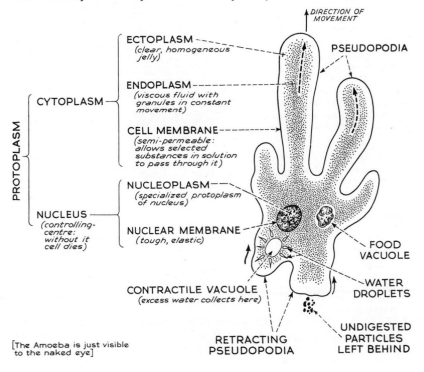

DIRECTION OF MOVEMENT

PROTOPLASM

CYTOPLASM

ECTOPLASM
(clear, homogeneous jelly)

PSEUDOPODIA

ENDOPLASM
(viscous fluid with granules in constant movement)

CELL MEMBRANE
(semi-permeable: allows selected substances in solution to pass through it)

NUCLEUS
(controlling-centre: without it cell dies)

NUCLEOPLASM
(specialized protoplasm of nucleus)

NUCLEAR MEMBRANE
(tough, elastic)

FOOD VACUOLE

WATER DROPLETS

CONTRACTILE VACUOLE
(excess water collects here)

UNDIGESTED PARTICLES LEFT BEHIND

RETRACTING PSEUDOPODIA

[The Amoeba is just visible to the naked eye]

1

The PHENOMENA which characterize all living things ---- are shown by the AMOEBA

① **ORGANIZATION**
Autoregulation —
inherent ability to control
all life processes.

② **IRRITABILITY**
Ability to respond to
stimuli (from changes
in the environment).

③ **CONTRACTILITY**
Ability to move.

④ **NUTRITION**
Ability to ingest,
digest, absorb
and assimilate
food.

⑤ **METABOLISM and GROWTH**
Ability to liberate potential
energy of food and to convert
it into mechanical work
(*e.g. movement*) and to
rebuild simple absorbed
units into the complex
protoplasm of the living cell.

⑥ **RESPIRATION**
Ability to take in oxygen
for oxidation of food
with release of energy;
and to eliminate the
resulting carbon
dioxide.

⑦ **EXCRETION**
Ability to eliminate
waste products of
metabolism.

⑧ **REPRODUCTION**
Ability to reproduce the
species.

PSEUDOPODIA FORMATION

Small elevation arises on surface

Cytoplasm streams forward

...forming long pseudopodium

Free floating

Contact made

Crawling along surface

Attracted by favourable stimuli.
Repelled by unfavourable stimuli.

FOOD VACUOLE FORMATION

INGESTION: *Amoeba flows round and engulfs food particle.*

SECRETION of Enzymes
(*chemical agents*) which
DIGEST (*break down*)
complex foods.

ABSORPTION &
UTILIZATION of
simple units by
the living cell.

EXPULSION
of
undigested
particles.

RESPIRATION and EXCRETION

Oxygen
(*in solution*)

Carbon
Dioxide
(*in solution*)

Waste Products
(*in solution*)

Water periodically
expelled at
surface

REPRODUCTION
Asexual in the amoeba
— by simple fission.

2

The CELL

Higher animals, including MAN, are made up of millions of living cells which vary widely in structure and function but have certain features in common.

A GENERALIZED ANIMAL CELL *(of secretory type)*. [The finer details are seen only in the high magnification afforded by ELECTRON MICROSCOPE.

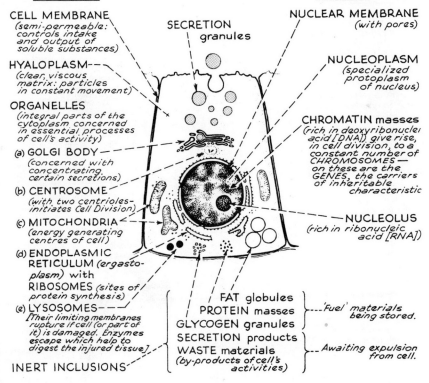

CYTOPLASM

CELL MEMBRANE
(semi-permeable: controls intake and output of soluble substances)

HYALOPLASM––
(clear, viscous matrix: particles in constant movement)

ORGANELLES
(integral parts of the cytoplasm concerned in essential processes of cell's activity)

(a) GOLGI BODY—
(concerned with concentrating certain secretions)

(b) CENTROSOME
(with two centrioles-initiates Cell Division)

(c) MITOCHONDRIA
(energy generating centres of cell)

(d) ENDOPLASMIC RETICULUM *(ergasto-plasm)* with RIBOSOMES *(sites of protein synthesis)*

(e) LYSOSOMES––
[Their limiting membranes rupture if cell (or part of it) is damaged. Enzymes escape which help to digest the injured tissue]

INERT INCLUSIONS

SECRETION granules

NUCLEUS

NUCLEAR MEMBRANE
(with pores)

NUCLEOPLASM
(specialized protoplasm of nucleus)

CHROMATIN masses
(rich in deoxyribonucleic acid [DNA]) give rise, in cell division, to a constant number of CHROMOSOMES — on these are the GENES, the carriers of inheritable characteristic

NUCLEOLUS
(rich in ribonucleic acid [RNA])

FAT globules
PROTEIN masses
GLYCOGEN granules }----'Fuel' materials being stored.

SECRETION products
WASTE materials *(by-products of cell's activities)* }-- Awaiting expulsion from cell.

3

CELL DIVISION *(MITOSIS)*

All cells arise from the division of pre-existing cells. In MITOSIS there is an exact QUALITATIVE division of the NUCLEUS and a less exact QUANTITATIVE division of the CYTOPLASM.

① INTERPHASE
Cell grows.
When contents double mitosis begins.

② PROPHASE I
Centrosome divides.
Centrioles migrate to opposite poles.
Chromatin appears as tangled skein of chromosomes.

③ PROPHASE II
Fine rays form spindle on which chromosomes align themselves.
Nucleolus and nuclear membrane disappear.

④ METAPHASE
Chromosomes jostle around equator of cell.

⑤ ANAPHASE I
Each chromosome splits lengthwise into two chromatids which now separate.

⑨ FINAL SEPARATION OF TWO NEW CELLS
Each carries the same units of heredity as the parent cell.

⑧ TELOPHASE II
Constriction deepens.
New nuclear membranes form.
Nucleoli appear.
Cell surface bubbles with activity as cells separate.

⑦ TELOPHASE I
Cytoplasm constricts.
Chromosomes at each pole come together and begin to fade.

⑥ ANAPHASE II
The separated chromatids move towards opposite poles of spindle.
Centrioles divide to form new centrosomes.

[For clarity only 4 chromosomes (2 pairs) are shown in these diagrams.]

The longitudinal halving of chromosomes and genes ensures that each new cell receives the same hereditary factors as the original cell.

The number of chromosomes is constant for any one species.
*The cells of the human body (somatic cells) carry 23 pairs —
i.e. 46 chromosomes. A modified division (Meiosis) gives 23
chromosomes in ova and sperm. Fertilization restores the 46.*

4

DIFFERENTIATION of ANIMAL CELLS

<u>SPECIALIZATION</u> distinguishes multicellular creatures from more primitive forms of life.

<u>ONE-CELLED ANIMALS</u> are capable of INDEPENDENT existence — they are *Undifferentiated* i.e. Show all activities or ⸺PHENOMENA of LIFE

<u>MANY-CELLED ANIMALS</u>   Cells CO-OPERATE for well-being of whole body.

Differentiated ⸺ Groups of cells undergo adaptations and sacrifice some powers to fit them for special duties.

All cells retain powers of
ORGANIZATION
IRRITABILITY
NUTRITION
METABOLISM
RESPIRATION
EXCRETION

<u>MODIFICATION</u>   *for*   <u>SPECIALIZATION</u>   *with*   <u>LOSS or REDUCTION</u>
<u>of STRUCTURE</u> *efficient*   <u>of FUNCTION</u>   <u>of VERSATILITY</u>
E.g.

SECRETORY CELL ⸺⸺ Highly developed powers of SECRETION e.g. enzymes for chemical breakdown of foodstuffs.

Cytoplasm   Nucleus displaced to base by formed and stored secretion

Nucleus

Diminished powers of CONTRACTION and REPRODUCTION

FAT CELL ⸺⸺ STORAGE of FAT

Cytoplasm   Cytoplasm displaced by stored fat

Nucleus

Loss of powers of CONTRACTION and SECRETION

MUSCLE CELL ⸺⸺ Highly developed powers of CONTRACTILITY

Cytoplasm   *Nucleus*

Elongated cell body

Diminished powers of SECRETION and REPRODUCTION

NERVE CELL ⸺⸺ Highly developed powers of IRRITABILITY
Cytoplasm

Cytoplasm drawn out into long branching processes

Nucleus
x400

(response to stimuli and transmission of impulses over long distances)

Loss of powers of REPRODUCTION

i.e. if nerve cell is destroyed no regeneration is possible.

ORGANIZATION OF TISSUES

Different cell types are not mixed haphazardly in the body.
Cells which are alike are arranged together to form TISSUES.
There are *four* main types of tissue:- 1. EPITHELIA or LINING,
2. CONNECTIVE or SUPPORTING, 3. MUSCULAR, 4. NERVOUS.

EPITHELIA

STRUCTURAL MODIFICATIONS	SITE	SPECIALIZED FUNCTIONS
Sheets of cells with minimum intercellular substance		*Line all internal and external surfaces of body*

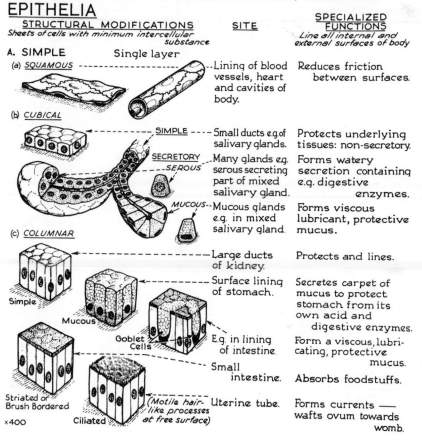

A. SIMPLE Single layer

(a) *SQUAMOUS* — Lining of blood vessels, heart and cavities of body. — Reduces friction between surfaces.

(b) *CUBICAL*

SIMPLE — Small ducts e.g. of salivary glands. — Protects underlying tissues: non-secretory.

SECRETORY SEROUS — Many glands e.g. serous secreting part of mixed salivary gland. — Forms watery secretion containing e.g. digestive enzymes.

MUCOUS — Mucous glands e.g. in mixed salivary gland. — Forms viscous lubricant, protective mucus.

(c) *COLUMNAR*

Large ducts of kidney. — Protects and lines.

Simple

Surface lining of stomach. — Secretes carpet of mucus to protect stomach from its own acid and digestive enzymes.

Mucous

Goblet Cells — E.g. in lining of intestine. — Form a viscous, lubricating, protective mucus.

Small intestine. — Absorbs foodstuffs.

Striated or Brush Bordered

×400

Ciliated — (Motile hair-like processes at free surface) Uterine tube. — Forms currents — wafts ovum towards womb.

6

B. PSEUDOSTRATIFIED COLUMNAR
CILIATED *(with Goblet cells)* ------- Respiratory passages.

(a) Protective.

(b) Cilia form currents to move mucus-trapped particles towards mouth.

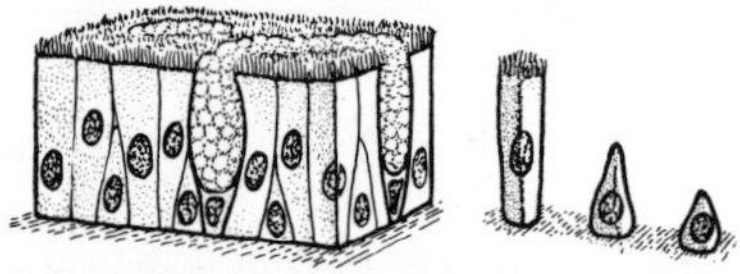

C. COMPOUND - Many layers of cells
TRUE STRATIFIED
(a) *TRANSITIONAL* --------------------- Urinary passages.

(a) Extensible.

(b) Protective — prevents penetration of urine into underlying tissues.

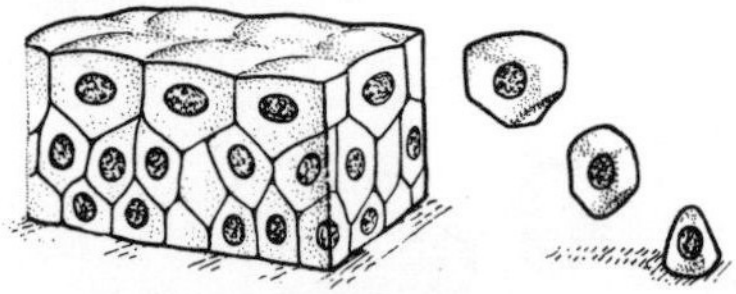

(b) *STRATIFIED SQUAMOUS* -------------- Surfaces subjected to great wear and tear.

Most resistant type of epithelium.

(i) *Uncornified.* ------- Internal surfaces e.g. mouth and gullet.

Protects against friction.

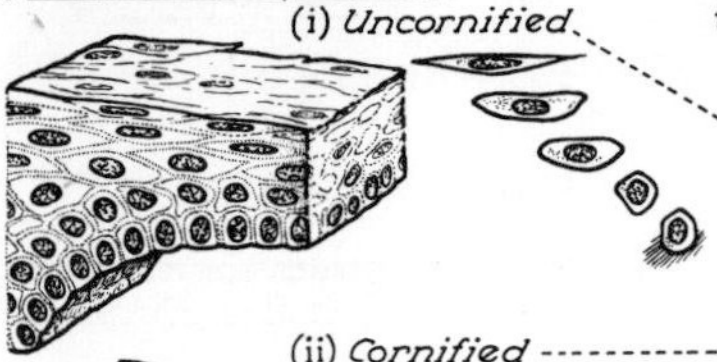

(ii) *Cornified* ---------- External skin surfaces.

Protects against wear and tear, evaporation and extremes of temperature.

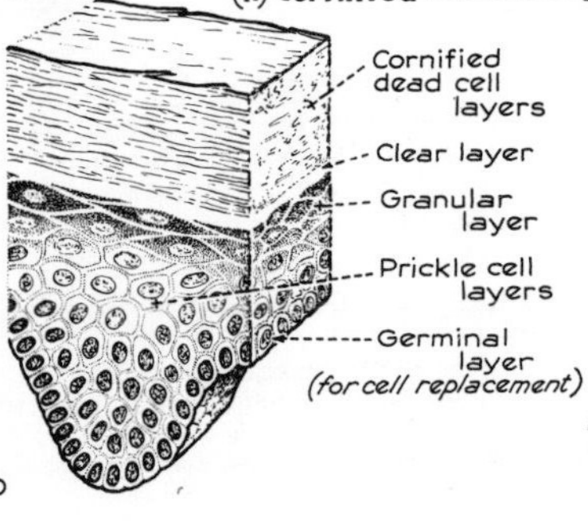

- Cornified dead cell layers
- Clear layer
- Granular layer
- Prickle cell layers
- Germinal layer *(for cell replacement)*

[Dead cell layers vary in thickness — e.g. thickest on soles of feet and palms of hands.]

×400

7

CONNECTIVE TISSUES

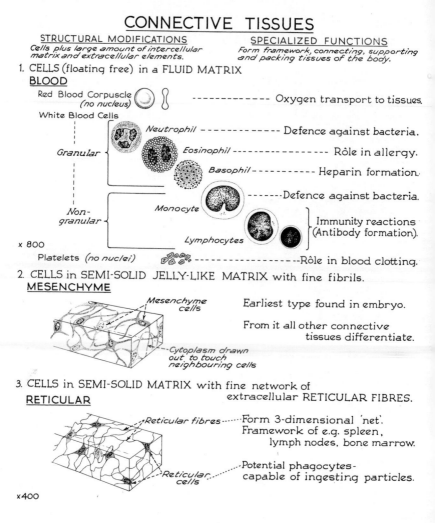

1. CELLS (floating free) in a FLUID MATRIX
   BLOOD

   Red Blood Corpuscle *(no nucleus)* ------------ Oxygen transport to tissues.

   White Blood Cells

   Granular

   Neutrophil ------------- Defence against bacteria.

   Eosinophil ----------------- Rôle in allergy.

   Basophil ----------- Heparin formation.

   ------- Defence against bacteria.

   Non-granular

   Monocyte

   } Immunity reactions (Antibody formation).

   Lymphocytes

   x 800

   Platelets *(no nuclei)* ------------------- Rôle in blood clotting.

2. CELLS in SEMI-SOLID JELLY-LIKE MATRIX with fine fibrils.
   MESENCHYME

   Mesenchyme cells

   Earliest type found in embryo.

   From it all other connective tissues differentiate.

   Cytoplasm drawn out to touch neighbouring cells

3. CELLS in SEMI-SOLID MATRIX with fine network of
   RETICULAR   extracellular RETICULAR FIBRES.

   Reticular fibres ------ Form 3-dimensional 'net'.
   Framework of e.g. spleen, lymph nodes, bone marrow.

   Reticular cells ------ Potential phagocytes-capable of ingesting particles.

   x400

8

CONNECTIVE TISSUES

4. CELLS in SEMI-SOLID MATRIX with thicker collagenous and elastic fibres.

(a) LOOSE FIBROUS

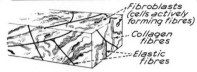

Fibroblasts (cells actively forming fibres)

Collagen fibres

Elastic fibres

"Packing" tissue between organs: sheaths of muscles, nerves and blood vessels.

(b) DENSE FIBROUS

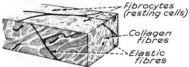

Fibrocytes (resting cells)

Collagen fibres

Elastic fibres

Strong, inelastic yet pliable — e.g. ligaments; capsules of joints; heart valves.

(c) ELASTIC

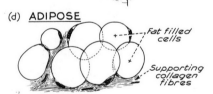

Elastic fibres (or laminae)

Strong, extensible and flexible — e.g. in walls of blood vessels and air passages.

(d) ADIPOSE

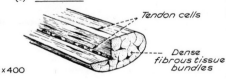

Fat filled cells

Supporting collagen fibres

Protective "cushion" for organs. Insulating layer in skin. Storage of fat reserves.

(e) TENDON

Tendon cells

Dense fibrous tissue bundles

Tough, inelastic cords of dense fibrous tissue which attach muscles to bones.

×400

9

CONNECTIVE TISSUES

5. CELLS in SOLID ELASTIC MATRIX with fibres.

(a) HYALINE CARTILAGE

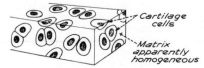

Firm yet resilient
e.g. in air passages; ends of bones at joints.
Transition stage in bone
formation.

Cartilage cells

Matrix apparently homogeneous

(b) WHITE FIBRO-CARTILAGE

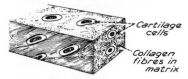

Tough. Resistant to stretching.
Acts as shock absorber between vertebrae.

Cartilage cells

Collagen fibres in matrix

(c) ELASTIC or YELLOW FIBRO-CARTILAGE

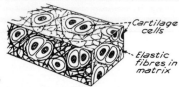

More flexible, resilient —
e.g. in larynx and ear.

Cartilage cells

Elastic fibres in matrix

×400

6. CELLS in SOLID RIGID MATRIX impregnated with Calcium and Magnesium Salts.

BONE

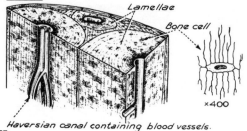

Lamellae

Bone cell

Forms rigid framework of body.

×400

Haversian canal containing blood vessels.

× 75

MUSCULAR TISSUES

All have ELONGATED cells with special development of CONTRACTILITY.

1. SMOOTH, UNSTRIPED, VISCERAL or INVOLUNTARY muscle

Nucleus

Least specialized. Shows slow rhythmical contraction and relaxation.
Not under voluntary control.
Found in walls of viscera and blood vessels.

2. CARDIAC or HEART muscle

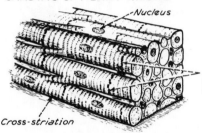

Nucleus

Cross-striation

More highly specialized.
Rapid rhythmical contraction (and relaxation) spreads through whole muscle mass.
Not under voluntary control.
Found only in heart wall.

Cells adhere, end to end, at intercalated discs to form long 'fibres' which branch and connect with adjacent 'fibres'.

(Note: There is no protoplasmic continuity between cells.)

3. STRIATED, STRIPED, SKELETAL or VOLUNTARY muscle

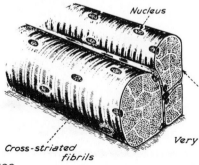

Nucleus

Cross-striated fibrils

×400

Most highly specialized.
Very rapid, powerful contractions of individual fibres.
Under voluntary control.
Found in e.g. muscles of trunk, limbs, head.

Thick covering membrane (sarcolemma)

Very long, large multi-nucleated units. No branching.

NERVOUS TISSUES

Nervous tissue is divided into:-
(a) **NEURONES** or *True* nerve
cells specialized in
IRRITABILITY, CONDUCTION,
INTEGRATION

MOTOR – pass messages from brain and spinal
cord to muscles and glands.
ASSOCIATION – relay messages between neurones.
SENSORY – receive and pass messages from
environment to brain and spinal cord.

(b) *Supporting* cells e.g. NEUROGLIA in brain and spinal cord;
SHEATH (*Schwann*) cells on nerve fibres outside brain and spinal cord.

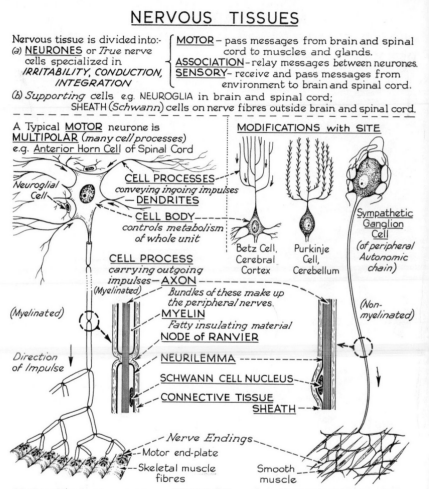

A Typical **MOTOR** neurone is
MULTIPOLAR (*many cell processes*)
e.g. <u>Anterior Horn Cell</u> of Spinal Cord

*Neuroglial
Cell*

MODIFICATIONS with SITE

CELL PROCESSES
conveying ingoing impulses
— DENDRITES

CELL BODY
controls metabolism
of whole unit

Betz Cell,
Cerebral
Cortex

Purkinje
Cell,
Cerebellum

Sympathetic
Ganglion
Cell
(*of peripheral
Autonomic
chain*)

CELL PROCESS
*carrying outgoing
impulses* — AXON

(*Myelinated*)

*Bundles of these make up
the peripheral nerves.*
MYELIN
Fatty insulating material
NODE of RANVIER

(*Myelinated*)

NEURILEMMA

(*Non-
myelinated*)

*Direction
of Impulse*

SCHWANN CELL NUCLEUS

CONNECTIVE TISSUE
SHEATH

- - *Nerve Endings* - - -
- - Motor end-plate
- - Skeletal muscle
fibres

Smooth
muscle

Most multipolar neurones are MOTOR (efferent) or ASSOCIATION in function.

NERVOUS TISSUES

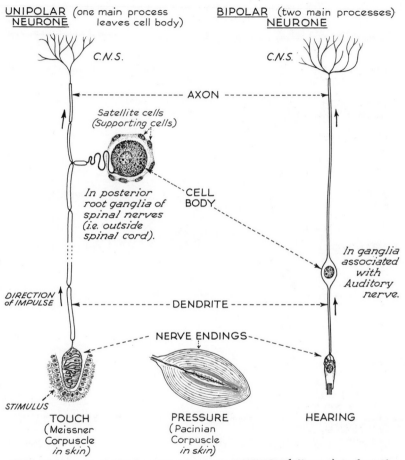

UNIPOLAR (one main process leaves cell body) NEURONE

BIPOLAR (two main processes) NEURONE

C.N.S.

C.N.S.

------- AXON -------

Satellite cells (Supporting cells)

In posterior root ganglia of spinal nerves (i.e. outside spinal cord).

CELL BODY

In ganglia associated with Auditory nerve.

DIRECTION of IMPULSE

------- DENDRITE -------

NERVE ENDINGS

STIMULUS

TOUCH (Meissner Corpuscle in skin)

PRESSURE (Pacinian Corpuscle in skin)

HEARING

All unipolar and bipolar neurones are SENSORY (afferent) in function.

2

The BODY SYSTEMS

The tissues are arranged to form ORGANS.
Organs are grouped into SYSTEMS.

The <u>ESSENTIAL LIFE</u> ---- are delegated to-------- <u>SEPARATE</u>
    <u>PROCESSES</u>   <u>SYSTEMS</u>

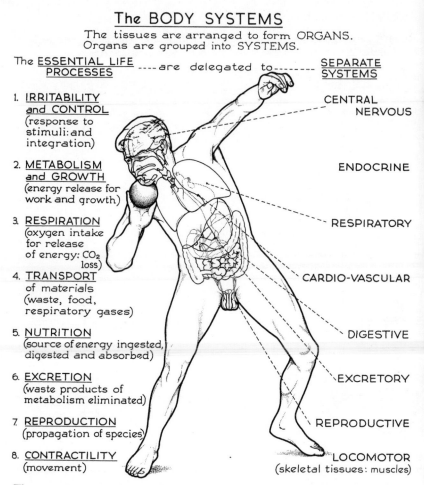

1. <u>IRRITABILITY</u>
<u>and CONTROL</u>
(response to
stimuli: and
integration)

CENTRAL
   NERVOUS

2. <u>METABOLISM</u>
<u>and GROWTH</u>
(energy release for
work and growth)

ENDOCRINE

3. <u>RESPIRATION</u>
(oxygen intake
for release
of energy: CO_2
     loss)

RESPIRATORY

4. <u>TRANSPORT</u>
of materials
(waste, food,
respiratory gases)

CARDIO-VASCULAR

5. <u>NUTRITION</u>
(source of energy ingested,
digested and absorbed)

DIGESTIVE

6. <u>EXCRETION</u>
(waste products of
metabolism eliminated)

EXCRETORY

7. <u>REPRODUCTION</u>
(propagation of species)

REPRODUCTIVE

8. <u>CONTRACTILITY</u>
(movement)

LOCOMOTOR
(skeletal tissues: muscles)

*The systems do not work independently. The body works as a whole.
Health and well-being depend on the co-ordinated effort of every part.*

CHAPTER 2.

NUTRITION, METABOLISM AND DIGESTION

BASIC CONSTITUENTS of PROTOPLASM

Protoplasm is made up of certain <u>Elements</u> present mainly in <u>Chemical Combination</u>

Eg. in a 70kg man these four elements make up most of the body weight:-

HYDROGEN(H) 6,580 g ⎤ A large amount of   ⎤ C,H and O combine   ⎤ C,H,O and N
                     ⎥ the H and O is   ⎥ chemically to form   ⎥ combine to form
OXYGEN (O) 43,550 g ⎦ present as WATER   ⎥ CARBOHYDRATES   ⎥ PROTEINS

CARBON (C) 12,590 g   ⎥ and FATS *(chief   ⎥ *(the main
                                              ⎥ sources of ENERGY   ⎥ BUILDING
NITROGEN (N) 1,815 g   ⎦ in living protoplasm)* ⎦ constituents of all protoplasm)*

The following eight make up much of the remaining body weight:-

CALCIUM (Ca) 1,700g ⎤ Important constituents of blood and of hard tissues-
PHOSPHORUS(P) 680g ⎦   e.g. bones and teeth.

CHLORINE (Cl)   115g ⎤ Important constituents of body fluids.
SODIUM (Na)   70g ⎦

POTASSIUM (K)   70g   Important constituent of all cells.
SULPHUR (S)   100g

MAGNESIUM (Mg)   42g   Important for activity of brain, nerves and muscles.
IRON (Fe)   7g   Important component of red blood cells.

Trace elements, such as manganese, copper, iodine, zinc, cobalt, fluorine, strontium, make up the last few grams or so.

Note:- *Apart from WATER, the chief constituents in both plant and animal protoplasm are present as compounds of CARBON, i.e. they are ORGANIC SUBSTANCES.*

<u>CARBOHYDRATES</u> — the simpler ones are SUGARS; the more complex— STARCH (made up of hundreds of units of sugar tied chemically together). They consist of atoms of C, H and O.

<u>FATS</u> — the molecule is made up of smaller molecules of FATTY ACIDS linked chemically with a molecule of GLYCEROL. They consist of atoms of the elements C, H and O.

<u>PROTEINS</u> — chief organic constituents of all protoplasm — the molecule is made up of smaller units called AMINO ACIDS. These consist of atoms of C, H, O, N and sometimes S and P.

There are thousands of different kinds of Protein but only about 20 different amino acids. Differences between proteins depend on the amino acids present and on their number and arrangement.

15

SOURCE of ENERGY: PHOTOSYNTHESIS

The essential life processes or the phenomena which characterize life depend on the use of ENERGY. The SUN is the SOURCE of ENERGY for ALL LIVING THINGS.

Only GREEN PLANTS can TRAP and STORE the sun's energy and build from relatively simple substances the energy-rich and body-building compounds — *CARBOHYDRATES, FATS and PROTEINS* — required by PROTOPLASM. This process is called <u>PHOTOSYNTHESIS</u>.

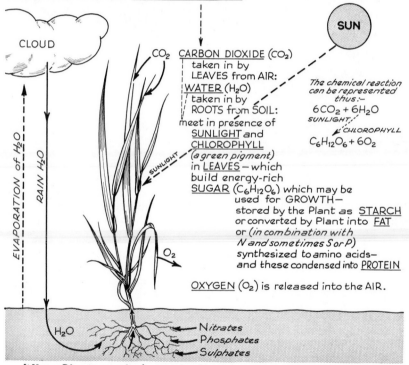

CLOUD

SUN

CO_2 <u>CARBON DIOXIDE</u> (CO_2)
taken in by
LEAVES from AIR:
<u>WATER</u> (H_2O)
taken in by
ROOTS from SOIL:
meet in presence of
<u>SUNLIGHT</u> and
<u>CHLOROPHYLL</u>
(a green pigment)
in <u>LEAVES</u> — which
build energy-rich
<u>SUGAR</u> ($C_6H_{12}O_6$) which may be
used for GROWTH—
stored by the Plant as <u>STARCH</u>
or converted by Plant into <u>FAT</u>
or *(in combination with
N and sometimes S or P)*
synthesized to amino acids—
and these condensed into <u>PROTEIN</u>

<u>OXYGEN</u> (O_2) is released into the AIR.

The chemical reaction can be represented thus :-

$$6CO_2 + 6H_2O$$
SUNLIGHT.·
CHLOROPHYLL
$$C_6H_{12}O_6 + 6O_2$$

EVAPORATION of H_2O

RAIN H_2O

SUNLIGHT

O_2

H_2O

Nitrates
Phosphates
Sulphates

When Plants or their products are eaten this stored energy becomes available to animals and man.

16

CARBON 'CYCLE' in NATURE

Animal bodies are unable to build *Proteins, Carbohydrates* or *Fats* from the raw materials. They must be built up for them by *Plants*.
CARBON is the basic building unit of all these compounds.

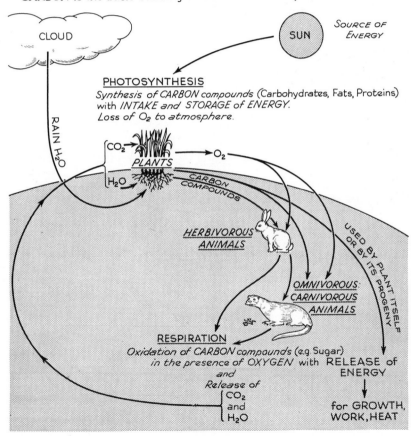

SOURCE OF ENERGY

CLOUD

SUN

RAIN H_2O

PHOTOSYNTHESIS
Synthesis of CARBON compounds (Carbohydrates, Fats, Proteins) *with INTAKE and STORAGE of ENERGY.*
Loss of O_2 to atmosphere.

CO_2
PLANTS
O_2
H_2O
CARBON COMPOUNDS

HERBIVOROUS ANIMALS

USED BY PLANT ITSELF OR BY ITS PROGENY

OMNIVOROUS: CARNIVOROUS ANIMALS

RESPIRATION
Oxidation of CARBON compounds (e.g. Sugar) *in the presence of OXYGEN with* RELEASE of ENERGY
and
Release of
CO_2
and
H_2O

for GROWTH, WORK, HEAT

NITROGEN 'CYCLE' in NATURE

Although animals are surrounded by NITROGEN in the air they can only use the NITROGEN trapped by Plants.

NITROGEN in the Atmosphere

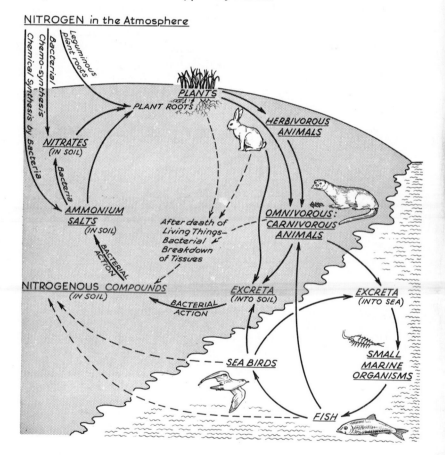

NUTRITION

Man eats as <u>FOOD</u> the substances which PLANTS
(and, through them, ANIMALS) have made.
These are broken down in man's body
into simpler chemical units which provide

<u>BUILDING and PROTECTIVE MATERIALS</u>

Man requires more or less the same
elements as plants. (*Some, such as
the minerals iodine, sodium or iron,
he assimilates in INORGANIC FORM.*)

Most must be built up for him
by plants :-
ORGANICALLY COMBINED
Carbon, Nitrogen, and Sulphur, etc.
ESSENTIAL AMINO ACIDS

ESSENTIAL FATTY ACIDS
Certain VITAMINS
These are
<u>USED to</u>

<u>BUILD, MAINTAIN or REPAIR</u>
PROTOPLASM

- - - - - - - - - - - - - - - - - - -

BODY BUILDING REQUIREMENTS
of the INDIVIDUAL determine
<u>QUALITY</u>
of DIET

- - - - - ENERGY

*Stored originally
by plants* <u>RELEASED</u> in man's cells
by
<u>OXIDATION</u>

*When food is 'burned' it gives up
its stored energy*

PROTEINS	yield 4·1 kilocalories		*Units of Energy per gram*
CARBOHYDRATES	yield 4·1 kilocalories		
FATS	yield 9·2 kilocalories		

Most of this appears as <u>HEAT</u>
and is <u>USED for</u>
<u>KEEPING BODY WARM</u>;
some is used for <u>WORK</u> of Cells

- - - - - - - - - - - - - - - - - - -

ENERGY REQUIREMENTS
of the INDIVIDUAL determine
<u>QUANTITY</u>
of DIET

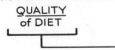

For a
<u>BALANCED DIET</u>
*TOTAL INTAKE
of*
Essential Constituents and Energy Units
must balance…
… Amounts *stored* plus amounts *lost* from body plus amounts *used*
as Work or Heat.

ENERGY-GIVING FOODS

All the main foodstuffs yield ENERGY – the energy originally trapped by Plants. CARBOHYDRATES and FATS are the chief energy-giving foods. PROTEINS can give energy but are mainly used for building and repairing protoplasm.

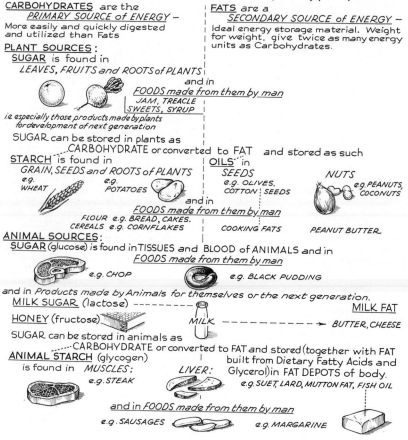

CARBOHYDRATES are the
  PRIMARY SOURCE of ENERGY –
More easily and quickly digested and utilized than Fats

PLANT SOURCES:
  SUGAR is found in
    LEAVES, FRUITS and ROOTS of PLANTS
                                and in
            FOODS made from them by man
                    JAM, TREACLE
                    SWEETS, SYRUP
i.e. especially those products made by plants for development of next generation

  SUGAR can be stored in plants as
          ......CARBOHYDRATE or converted to FAT and stored as such
  STARCH is found in
    GRAIN, SEEDS and ROOTS of PLANTS
      e.g.   e.g.
      WHEAT   POTATOES
                            and in
            FOODS made from them by man
        FLOUR e.g. BREAD, CAKES.
        CEREALS e.g. CORNFLAKES

FATS are a
  SECONDARY SOURCE of ENERGY –
Ideal energy storage material. Weight for weight, give twice as many energy units as Carbohydrates.

OILS in
  SEEDS   *NUTS*
  e.g. OLIVES,   e.g. PEANUTS,
  COTTON SEEDS   COCONUTS

  COOKING FATS   PEANUT BUTTER

ANIMAL SOURCES:
  SUGAR (glucose) is found in TISSUES and BLOOD of ANIMALS and in
              FOODS made from them by man
        e.g. CHOP   e.g. BLACK PUDDING

and in *Products made by Animals for themselves or the next generation.*
  MILK SUGAR (lactose) – – – – – – – –   MILK FAT
  HONEY (fructose)   MILK – – – – – – – – → BUTTER, CHEESE
  SUGAR can be stored in animals as
          ......CARBOHYDRATE or converted to FAT and stored (together with FAT
  ANIMAL STARCH (glycogen)   built from Dietary Fatty Acids and
    is found in *MUSCLES*:   *LIVER*:   Glycerol) in FAT DEPOTS of body.
        e.g. STEAK   e.g. SUET, LARD, MUTTON FAT, FISH OIL

              and in FOODS made from them by man
        e.g. SAUSAGES   e.g. MARGARINE

BODY–BUILDING FOODS

PROTEIN is the chief body-building food. Because it is the chief constituent of protoplasm, tissues of Plants and Animals are the richest sources.

PLANT SOURCES

Protoplasm of Plant Tissues and — *Stores made by Plants* — or *Foods made for the next generation in seeds and roots* *from them by man*.

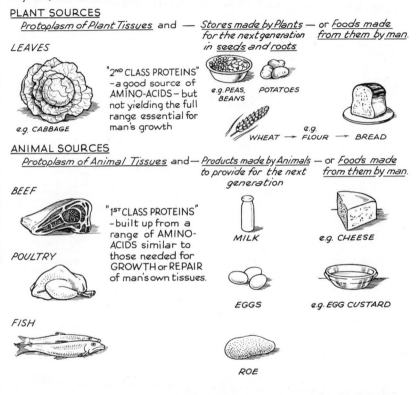

LEAVES

"2ND CLASS PROTEINS" – a good source of AMINO-ACIDS – but not yielding the full range essential for man's growth

e.g. PEAS, BEANS   *POTATOES*

e.g. CABBAGE

WHEAT → *FLOUR* → *BREAD* *e.g.*

ANIMAL SOURCES

Protoplasm of Animal Tissues and — *Products made by Animals* — or *Foods made to provide for the next generation* *from them by man*.

BEEF

"1ST CLASS PROTEINS" – built up from a range of AMINO-ACIDS similar to those needed for GROWTH or REPAIR of man's own tissues.

MILK

e.g. CHEESE

POULTRY

EGGS

e.g. EGG CUSTARD

FISH

ROE

These foods between them also contain other important body-building elements:– e.g. CALCIUM and PHOSPHORUS for making BONES and TEETH hard. IRON for building HAEMOGLOBIN– the OXYGEN CARRYING substance in RED BLOOD CELLS. IODINE for building the THYROID HORMONE.

PROTECTIVE FOODS — I

Even when provided with BODY-BUILDING and ENERGY-GIVING FOODS, human tissues cannot build or repair their protoplasm or release the energy in the absence of 'PROTECTIVE' substances called <u>VITAMINS</u>. These appear to be important links in METABOLIC processes. Vitamins cannot be substituted for one another in the way that different Carbohydrates, Proteins or Fats can replace others.

All are essential for NORMAL GROWTH, HEALTH and WELL-BEING and for general RESISTANCE to INFECTION.

EACH has additional specific protective functions:-

FAT-SOLUBLE VITAMINS	**A** ANTIXEROPH-THALMIC	**D** ANTI-RACHITIC	**E** ANTISTERILITY	**K** ANTI-HAEMORRHAGIC
	Protects skin and mucous membranes (e.g. digestive, respiratory) and is essential for regeneration of Visual Purple in eye.	Essential for deposition of Ca and P in bones and teeth. Promotes absorption of Ca from intestine.	Perhaps important for normal reproduction in adult.	Important for normal blood clotting.
PLANT SOURCES LEAVES— SEEDS & FRUITS— ROOTS—	Present as precursor CAROTENE. Green, yellow veg. (esp. spinach & kale) Yellow maize, peas, beans CARROTS	Vegetables, fruits and cereals contain negligible amounts	Green leaves (e.g. lettuce), peas. Richest source-Germ of various cereals e.g. WHEAT GERM OIL.	Spinach, kale, cabbage, cauliflower, cereals, tomatoes, carrots, potatoes.
ANIMAL SOURCES TISSUES— PRODUCTS— for PROGENY FOODS made from them	Animal liver breaks down CAROTENE to Vit.A Stored in LIVER of animals and FISH Secreted in MILK, EGG YOLK. BUTTER. CREAM.	Formed when ultra-violet rays fall on sterols in man's skin. Stored in LIVER of FISH and animals (COD LIVER OIL) MILK EGGS BUTTER Vitamin con- CREAM tent varies CHEESE with season.	Small amounts in Meat and Dairy Produce	Synthesized by BACTERIA in man's INTESTINE then absorbed in presence of BILE.
DIETARY DEFICIENCY *of any or all vitamins retards* GROWTH *and leads to* SEVERE DISEASE *and eventually* DEATH	XEROPH-THALMIA 	RICKETS *(in child)* OSTEOMALACIA *(in adult)*	Little evidence of deficiency in human beings. (In some animals developing embryos die. In male, testes may show degenerative changes.)	If deficient absorption from intestine → upset in mechanism for Blood Clotting

PROTECTIVE FOODS — II

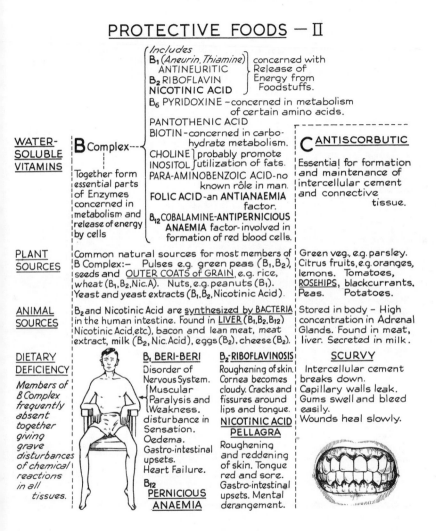

	B Complex	**C ANTISCORBUTIC**
WATER-SOLUBLE VITAMINS	*Includes* B₁ (*Aneurin, Thiamine*) ANTINEURITIC B₂ RIBOFLAVIN NICOTINIC ACID } concerned with Release of Energy from Foodstuffs. B₆ PYRIDOXINE – concerned in metabolism of certain amino acids. PANTOTHENIC ACID BIOTIN – concerned in carbohydrate metabolism. CHOLINE] probably promote INOSITOL } utilization of fats. PARA-AMINOBENZOIC ACID – no known rôle in man. FOLIC ACID – an ANTIANAEMIA factor. B₁₂ COBALAMINE - ANTIPERNICIOUS ANAEMIA factor - involved in formation of red blood cells.	Essential for formation and maintenance of intercellular cement and connective tissue.
	Together form essential parts of Enzymes concerned in metabolism and release of energy by cells	

Water-soluble vitamins — B Complex: Together form essential parts of Enzymes concerned in metabolism and release of energy by cells.

The B Complex includes:
- B₁ (*Aneurin, Thiamine*) ANTINEURITIC
- B₂ RIBOFLAVIN
- NICOTINIC ACID — concerned with Release of Energy from Foodstuffs.
- B₆ PYRIDOXINE – concerned in metabolism of certain amino acids.
- PANTOTHENIC ACID
- BIOTIN – concerned in carbohydrate metabolism.
- CHOLINE] probably promote utilization of fats.
- INOSITOL]
- PARA-AMINOBENZOIC ACID – no known rôle in man.
- FOLIC ACID – an ANTIANAEMIA factor.
- B₁₂ COBALAMINE - ANTIPERNICIOUS ANAEMIA factor - involved in formation of red blood cells.

C ANTISCORBUTIC: Essential for formation and maintenance of intercellular cement and connective tissue.

PLANT SOURCES

Common natural sources for most members of B Complex:– Pulses e.g. green peas (B₁, B₂), seeds and OUTER COATS OF GRAIN, e.g. rice, wheat (B₁, B₂, Nic.A). Nuts, e.g. peanuts (B₁). Yeast and yeast extracts (B₁, B₂, Nicotinic Acid).

Green veg., e.g. parsley. Citrus fruits, e.g. oranges, lemons. Tomatoes, ROSEHIPS, blackcurrants. Peas. Potatoes.

ANIMAL SOURCES

B₂ and Nicotinic Acid are synthesized by BACTERIA in the human intestine. Found in LIVER (B₁, B₂, B₁₂, Nicotinic Acid, etc.), bacon and lean meat, meat extract, milk (B₂, Nic.Acid), eggs (B₂), cheese (B₂).

Stored in body – High concentration in Adrenal Glands. Found in meat, liver. Secreted in milk.

DIETARY DEFICIENCY

Members of B Complex frequently absent together giving grave disturbances of chemical reactions in all tissues.

B₁ BERI-BERI
Disorder of Nervous System. { Muscular Paralysis and Weakness. } disturbance in Sensation. Oedema. Gastro-intestinal upsets. Heart Failure.

B₁₂ PERNICIOUS ANAEMIA

B₂ RIBOFLAVINOSIS
Roughening of skin. Cornea becomes cloudy. Cracks and fissures around lips and tongue.

NICOTINIC ACID PELLAGRA
Roughening and reddening of skin. Tongue red and sore. Gastro-intestinal upsets. Mental derangement.

SCURVY
Intercellular cement breaks down. Capillary walls leak. Gums swell and bleed easily. Wounds heal slowly.

23

ENERGY REQUIREMENTS *MALE*

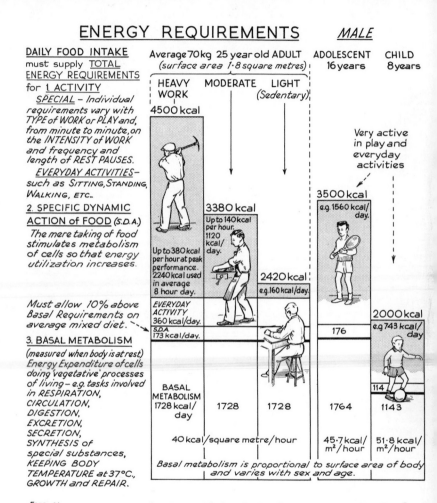

DAILY FOOD INTAKE must supply TOTAL ENERGY REQUIREMENTS for 1 ACTIVITY

SPECIAL – *Individual requirements vary with TYPE of WORK or PLAY and, from minute to minute, on the INTENSITY of WORK and frequency and length of REST PAUSES.*

EVERYDAY ACTIVITIES– *such as SITTING, STANDING, WALKING, ETC.*

2. SPECIFIC DYNAMIC ACTION of FOOD (S.D.A.)

The mere taking of food stimulates metabolism of cells so that energy utilization increases.

Must allow 10% above Basal Requirements on average mixed diet.

3. BASAL METABOLISM

(measured when body is at rest) *Energy Expenditure of cells doing 'vegetative' processes of living – e.g. tasks involved in RESPIRATION, CIRCULATION, DIGESTION, EXCRETION, SECRETION, SYNTHESIS of special substances, KEEPING BODY TEMPERATURE at 37°C, GROWTH and REPAIR.*

Average 70 kg 25 year old ADULT (surface area 1·8 square metres) — ADOLESCENT 16 years — CHILD 8 years

	HEAVY WORK	MODERATE	LIGHT (Sedentary)		
	4500 kcal				
					Very active in play and everyday activities
				3500 kcal e.g. 1560 kcal/day.	
	Up to 380 kcal per hour at peak performance. 2240 kcal used in average 8 hour day.	3380 kcal Up to 140 kcal per hour. 1120 kcal/day.	2420 kcal e.g. 160 kcal/day.		
EVERYDAY ACTIVITY	360 kcal/day.			176	2000 kcal e.g. 743 kcal/day
S.D.A.	173 kcal/day.				114
BASAL METABOLISM	1728 kcal/day	1728	1728	1764	1143
	40 kcal/square metre/hour			45·7 kcal/m²/hour	51·8 kcal/m²/hour

Basal metabolism is proportional to surface area of body and varies with sex and age.

[All figures are approximate and intended only as a general guide.]

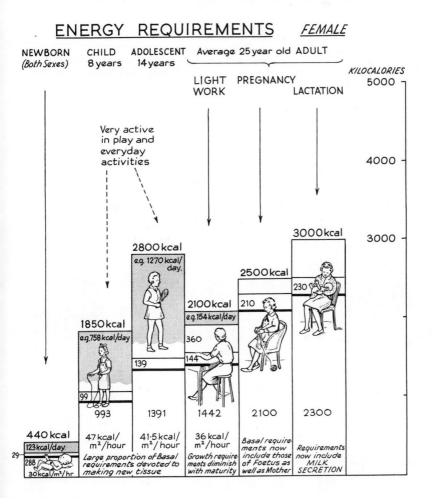

ENERGY REQUIREMENTS *FEMALE*

NEWBORN (Both Sexes) — CHILD 8 years — ADOLESCENT 14 years — Average 25 year old ADULT

KILOCALORIES

LIGHT WORK — PREGNANCY — LACTATION

Very active in play and everyday activities

5000

4000

3000

3000 kcal
230

2500 kcal
210

2800 kcal
e.g. 1270 kcal/day.
139

2100 kcal
e.g. 154 kcal/day
360
144

1850 kcal
e.g. 758 kcal/day
99

440 kcal
123 kcal/day
29
288
30 kcal/m²/hr

993 — 1391 — 1442 — 2100 — 2300

	47 kcal/ m²/hour	41·5 kcal/ m²/hour	36 kcal/ m²/hour		
	Large proportion of Basal requirements devoted to making new tissue		*Growth requirements diminish with maturity*	*Basal requirements now include those of Foetus as well as Mother*	*Requirements now include MILK SECRETION*

[Proportion needed for growth diminishes in both sexes with age.]

25

BALANCED DIET

The individual's daily energy requirements are best obtained by eating well-balanced meals which contain the essential *body-building* and *protective* foods as well as *energy-giving* foods.

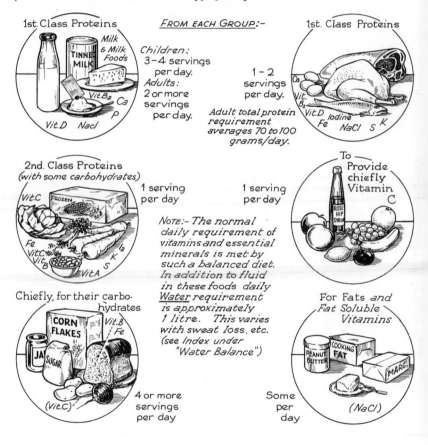

1st. Class Proteins

Milk & Milk Foods

TINNED MILK

Vit.B₂ — Ca — P — Vit.D — Nacl

FROM EACH GROUP:—

Children:
3–4 servings per day.
Adults:
2 or more servings per day.

1 – 2 servings per day.

Adult total protein requirement averages 70 to 100 grams/day.

1st. Class Proteins

Ca — Vit.B₂ — Vit.D — Fe — Iodine — NaCl — S — K

2nd. Class Proteins
(with some carbohydrates)

Vit.C — FROZEN — Fe — Vit.C — Vit.B — Vit.A — K — S

1 serving per day

1 serving per day

NOTE:— The normal daily requirement of vitamins and essential minerals is met by such a balanced diet. In addition to fluid in these foods daily <u>Water</u> *requirement is approximately 1 litre. This varies with sweat loss, etc. (see Index under "Water Balance")*

To — Provide chiefly Vitamin C

ROSE HIP SYRUP

Chiefly, for their carbo-hydrates

CORN FLAKES — Vit.B — Fe — JAM — SUGAR — (Vit.C)

4 or more servings per day

For Fats *and* Fat Soluble *Vitamins*

PEANUT BUTTER — COOKING FAT — MARG.

Some per day

(NaCl)

DIGESTIVE SYSTEM

The **ALIMENTARY CANAL** and } Special System for dealing
      ASSOCIATED GLANDS } with <u>**FOOD and FLUIDS**</u>

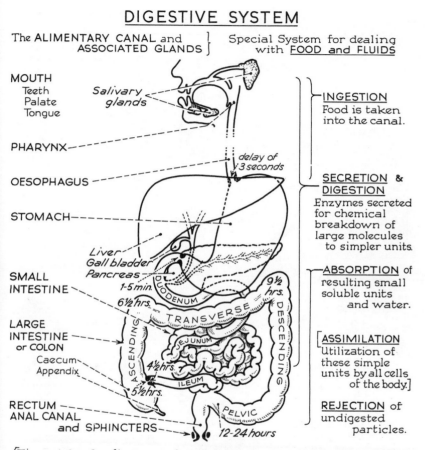

MOUTH
  Teeth
  Palate
  Tongue

Salivary glands

PHARYNX

OESOPHAGUS

delay of 3 seconds

STOMACH

Liver
Gall bladder
Pancreas

SMALL INTESTINE

1-5 min.

DUODENUM

6½ hrs.

9½ hrs.

LARGE INTESTINE or **COLON**
  Caecum
  Appendix

ASCENDING

TRANSVERSE

JEJUNUM

DESCENDING

4½ hrs.

ILEUM

5½ hrs.

RECTUM
ANAL CANAL
  and **SPHINCTERS**

PELVIC

12-24 hours

<u>**INGESTION**</u>
Food is taken
into the canal.

<u>**SECRETION**</u> &
<u>**DIGESTION**</u>
Enzymes secreted
for chemical
breakdown of
large molecules
to simpler units.

<u>**ABSORPTION**</u> of
resulting small
soluble units
    and water.

[<u>**ASSIMILATION**</u>
Utilization of
these simple
units by all cells
of the body.]

<u>**REJECTION**</u> of
undigested
    particles.

[Times taken by first part of meal to reach various parts of canal after leaving mouth are indicated. Latter part takes 3-5 hrs. to leave stomach.]

During its progress along the canal FOOD is subjected to MECHANICAL as well as CHEMICAL changes to make it suitable for absorption into blood stream.

27

DIGESTION in the MOUTH

MECHANICAL PROCESSES

MASTICATION

Chewing movements of TEETH, TONGUE, CHEEKS, LIPS, LOWER JAW, break down food, mix it with SALIVA and roll it into a moist soft mass (BOLUS) suitable for SWALLOWING

[Mastication is under VOLUNTARY control]

PAROTID SALIVARY GLAND (serous)

x200

SUBLINGUAL and SUBMANDIBULAR SALIVARY GLANDS (mixed mucous and serous glands)

Crescent of Serous cells

x200

CHEMICAL PROCESSES

SALIVA (1-1½ litres per day) is a slightly acid solution of salts and organic substances secreted mainly by 3 pairs of SALIVARY GLANDS.

SEROUS ACINI → Clear watery fluid containing salts and PTYALIN – an enzyme which starts to split cooked STARCH → DEXTRINS → MALTOSE (a simpler sugar)

WATER
Solid substances must be dissolved in saliva to stimulate TASTE buds.

MUCOUS ACINI → Thick slimy secretion of MUCIN → lubricant coating to food to assist SWALLOWING

[Salivary secretion is REFLEX and INVOLUNTARY]

Other important (non-digestive) functions of SALIVA :-

CLEANSING — Mouth and teeth kept free of debris, etc., to inhibit bacteria.

MOISTENING and LUBRICATING — Soft parts of mouth kept pliable for SPEECH.

EXCRETORY— Many organic substances (e.g. urea, sugar) and inorganic substances (e.g. mercury, lead) can be excreted in saliva.

28

CONTROL of SALIVARY SECRETION

Increased secretion at mealtimes is REFLEX *(involuntary)*.
Salivary Reflexes are of two types:-

(a) **UNCONDITIONED** *(inborn)*

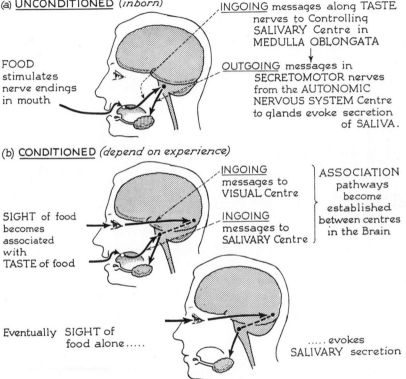

INGOING messages along TASTE nerves to Controlling SALIVARY Centre in MEDULLA OBLONGATA

FOOD stimulates nerve endings in mouth

OUTGOING messages in SECRETOMOTOR nerves from the AUTONOMIC NERVOUS SYSTEM Centre to glands evoke secretion of SALIVA.

(b) **CONDITIONED** *(depend on experience)*

SIGHT of food becomes associated with TASTE of food

INGOING messages to VISUAL Centre

INGOING messages to SALIVARY Centre

ASSOCIATION pathways become established between centres in the Brain

Eventually SIGHT of food alone......

..... evokes SALIVARY secretion

Similar conditioned reflexes are established by smell, by thought of food, and even by the sounds of its preparation.

3

OESOPHAGUS

The Oesophagus is a muscular tube about 25 cm. long which conveys ingested food and fluid from the Mouth to the Stomach.

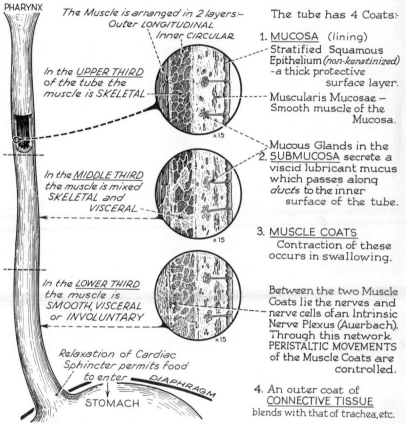

PHARYNX

The Muscle is arranged in 2 layers:—
Outer LONGITUDINAL
Inner CIRCULAR

In the UPPER THIRD of the tube the muscle is SKELETAL

In the MIDDLE THIRD the muscle is mixed SKELETAL and VISCERAL

In the LOWER THIRD the muscle is SMOOTH, VISCERAL or INVOLUNTARY

Relaxation of Cardiac Sphincter permits food to enter

DIAPHRAGM

STOMACH

×15

The tube has 4 Coats:—

1. MUCOSA (lining)
- Stratified Squamous Epithelium (non-keratinized) - a thick protective surface layer.
- Muscularis Mucosae — Smooth muscle of the Mucosa.

Mucous Glands in the
2. SUBMUCOSA secrete a viscid lubricant mucus which passes along ducts to the inner surface of the tube.

3. MUSCLE COATS
Contraction of these occurs in swallowing.

Between the two Muscle Coats lie the nerves and nerve cells of an Intrinsic Nerve Plexus (Auerbach). Through this network PERISTALTIC MOVEMENTS of the Muscle Coats are controlled.

4. An outer coat of CONNECTIVE TISSUE blends with that of trachea, etc.

Except during passage of food the oesophagus is flattened and closed; its mucosa thrown into several longitudinal folds.

SWALLOWING

Swallowing is a complex act initiated *voluntarily* and completed *involuntarily* (or *reflexly*)

STIMULUS | CENTRE IN MEDULLA OBLONGATA | EFFECT

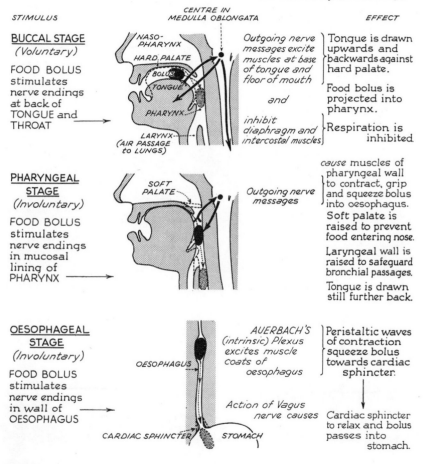

BUCCAL STAGE
(*Voluntary*)

FOOD BOLUS
stimulates
nerve endings
at back of
TONGUE and
THROAT

Outgoing nerve messages excite muscles at base of tongue and floor of mouth

and

inhibit diaphragm and intercostal muscles

Tongue is drawn upwards and backwards against hard palate.

Food bolus is projected into pharynx.

Respiration is inhibited

NASO-PHARYNX, HARD PALATE, BOLUS, TONGUE, PHARYNX, LARYNX (AIR PASSAGE TO LUNGS)

PHARYNGEAL STAGE
(*Involuntary*)

FOOD BOLUS
stimulates
nerve endings
in mucosal
lining of
PHARYNX

SOFT PALATE

Outgoing nerve messages

cause muscles of pharyngeal wall to contract, grip and squeeze bolus into oesophagus.

Soft palate is raised to prevent food entering nose.

Laryngeal wall is raised to safeguard bronchial passages.

Tongue is drawn still further back.

OESOPHAGEAL STAGE
(*Involuntary*)

FOOD BOLUS
stimulates
nerve endings
in wall of
OESOPHAGUS

OESOPHAGUS

CARDIAC SPHINCTER   STOMACH

AUERBACH'S (intrinsic) Plexus excites muscle coats of oesophagus

Action of Vagus nerve causes

Peristaltic waves of contraction squeeze bolus towards cardiac sphincter.

Cardiac sphincter to relax and bolus passes into stomach.

STOMACH

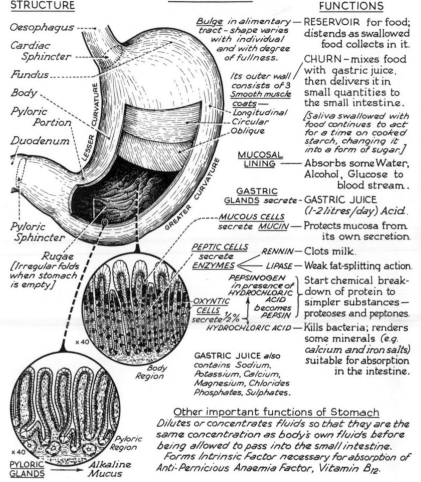

STRUCTURE

Oesophagus

Cardiac Sphincter

Fundus

Body

Pyloric Portion

Duodenum

LESSER CURVATURE

Pyloric Sphincter

Rugae
[Irregular folds when stomach is empty]

GREATER CURVATURE

x 40

Body Region

x 40

Pyloric Region

PYLORIC GLANDS ⟶ Alkaline Mucus

Bulge in alimentary tract – shape varies with individual and with degree of fullness.

Its outer wall consists of 3 Smooth muscle coats —
Longitudinal
Circular
Oblique

MUCOSAL LINING

GASTRIC GLANDS *secrete*

MUCOUS CELLS secrete MUCIN

PEPTIC CELLS secrete ENZYMES ⟵ RENNIN
LIPASE

PEPSINOGEN in presence of HYDROCHLORIC ACID becomes PEPSIN

OXYNTIC CELLS secrete ½%

HYDROCHLORIC ACID

GASTRIC JUICE *also contains Sodium, Potassium, Calcium, Magnesium, Chlorides Phosphates, Sulphates.*

FUNCTIONS

RESERVOIR for food; distends as swallowed food collects in it.

CHURN – mixes food with gastric juice, then delivers it in small quantities to the small intestine.

[Saliva swallowed with food continues to act for a time on cooked starch, changing it into a form of sugar.]

Absorbs some Water, Alcohol, Glucose to blood stream.

GASTRIC JUICE (1-2 litres/day) Acid.

Protects mucosa from its own secretion.

RENNIN — Clots milk.

LIPASE — Weak fat-splitting action.

Start chemical breakdown of protein to simpler substances — proteoses and peptones.

HYDROCHLORIC ACID — Kills bacteria; renders some minerals (e.g. calcium and iron salts) suitable for absorption in the intestine.

Other important functions of Stomach
Dilutes or concentrates fluids so that they are the same concentration as body's own fluids before being allowed to pass into the small intestine.
Forms Intrinsic Factor necessary for absorption of Anti-Pernicious Anaemia Factor, Vitamin B$_{12}$.

GASTRIC JUICE

SECRETION of Gastric juice is under two types of CONTROL:-

(a) NERVOUS – Messages are conveyed rapidly from Brain Centre by nerve fibres of the AUTONOMIC NERVOUS SYSTEM for *immediate* effect. *Eg.* stimulation of Parasympathetic (Vagus) nerves to GASTRIC GLANDS → secretion of HIGHLY ACID JUICE containing ENZYMES.

NERVOUS effects are brought about REFLEXLY:-

	STIMULUS	INGOING PATHWAY	
(i) Before eating	→ Thought, Sight, Smell of food	→ Association pathways in brain	→
(ii) While food is in mouth	→ Taste of food	→ Nerves of Taste	→
(iii) After swallowing food	→ Food distends stomach wall	→ Vagus	→

CENTRE IN MEDULLA

OUTGOING MESSAGES in

Vagus

Secretion of Gastric Juice

HEART   LUNGS   LIVER   STOMACH

(b) HUMORAL – Here the message is CHEMICAL and carried in the BLOOD STREAM for *slower and longer-lasting* control

Food in Stomach (*esp products of Protein digestion, extracts of meat*) stimulates Pyloric mucosa to secrete hormones

| 1. GASTRIN..... | *These travel in* Blood leaving stomach | → General Circulation | → *Return in* Arterial blood to stomach | 1. Secretion of juice rich in *Acid.* |
| 2. GASTROZYMIN.... | | | | 2. rich in *Enzymes.* |

When products of protein digestion reach duodenum	INTESTINAL GASTRIN is formed	→ Blood leaving duodenum	→ General Circulation	→ Arterial blood to stomach	→ Secretion of Gastric Juice.
Chyme (*especially fat*) accumulating in duodenum	ENTERO-GASTRONE	→ Blood leaving duodenum	→ General Circulation	→ Arterial blood to stomach	→ Inhibits secretion of Gastric Juice near end of meal.

33

MOVEMENTS of the STOMACH

Very little movement is seen in empty stomach until onset of HUNGER.

FILLING

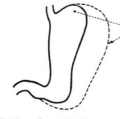

FUNDUS and
GREATER CURVATURE bulge and lengthen
as stomach fills with food. This is
"Receptive Relaxation"–the smooth muscle
cells relax so that the pressure in the
stomach does not rise until the organ is
fairly large.

TWO TYPES of MOVEMENT occur while food is in stomach.

TONUS

HOPPER

BODY – Food lies in layers and the walls
exert a slight but steady pressure on it.
This squeezes food steadily towards the
PYLORUS even when stomach is becoming
relatively empty near the end of a meal.

PERISTALSIS

MILL

PYLORIC
SPHINCTER

From INCISURA ANGULARIS vigorous waves
of contraction mix food with digestive
juices and carry chyme through the
normally relaxed PYLORUS into the first
part of the duodenum.

EMPTYING

CONTROL of MOTILITY and EMPTYING

1. Enterogastric
   NERVOUS Reflex   *inhibition of VAGUS→* | Inhibits peristalsis temporarily.
   As Chyme enters and | Gastric motility depressed.
   distends duodenum
   HUMORAL   *release of ENTEROGASTRONE→* | Temporarily slows emptying
                         to Blood Stream | of stomach.

2.
   As duodenum   *stimulation of VAGUS →* | Waves of peristalsis become
   empties | stronger.
                                              | Gastric motility increased.
   withdrawal of ENTEROGASTRONE→ | Emptying speeded up again.

*The hormone is particularly important in regulating the rate of emptying of the
stomach from moment to moment during the digestion of a meal.*

*[STARCHY FOODS leave the stomach quickly: MEAT leaves relatively slowly:
FATTY FOODS pass through most slowly of all.]*

PANCREAS

The Pancreas is a large gland lying across the Posterior Abdominal Wall. It has 2 secretions – a *digestive* secretion poured into the duodenum, and a *hormonal* secretion passed into the blood stream.

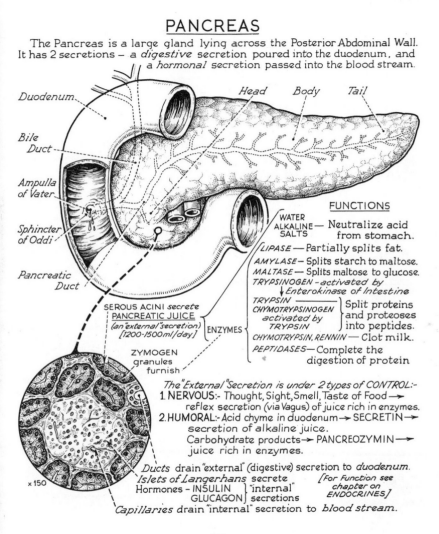

Duodenum

Head   Body   Tail

Bile Duct

Ampulla of Vater

Sphincter of Oddi

Pancreatic Duct

SEROUS ACINI *secrete*
PANCREATIC JUICE
(an "external" secretion)
[1200-1500ml/day]

ZYMOGEN granules furnish

× 150

FUNCTIONS

WATER
ALKALINE — Neutralize acid
SALTS   from stomach.

LIPASE — Partially splits fat.

AMYLASE – Splits starch to maltose.
MALTASE — Splits maltose to glucose.
TRYPSINOGEN – activated by
   ↓ *Enterokinase of Intestine*
TRYPSIN }
CHYMOTRYPSINOGEN } Split proteins
  activated by } and proteoses
  TRYPSIN } into peptides.
CHYMOTRYPSIN, RENNIN — Clot milk.
PEPTIDASES — Complete the digestion of protein

ENZYMES

The "External" Secretion is under 2 types of CONTROL:-
1. NERVOUS:- Thought, Sight, Smell, Taste of Food →
   reflex secretion (via Vagus) of juice rich in enzymes.
2. HUMORAL:- Acid chyme in duodenum → SECRETIN →
   secretion of alkaline juice.
   Carbohydrate products → PANCREOZYMIN →
   juice rich in enzymes.

Ducts drain "external" (digestive) secretion to *duodenum*.
Islets of Langerhans secrete   [For Function see
Hormones – INSULIN } "internal"   chapter on
         GLUCAGON } secretions   ENDOCRINES]
Capillaries drain "internal" secretion to *blood stream*.

LIVER (and GALL BLADDER)

The Liver is a large, highly complex organ with many functions. One of these is to secrete 500-1000 ml. of BILE per day.

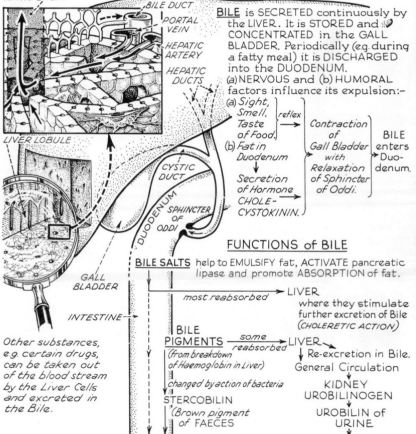

BILE is SECRETED continuously by the LIVER. It is STORED and CONCENTRATED in the GALL BLADDER. Periodically (eg. during a fatty meal) it is DISCHARGED into the DUODENUM. (a) NERVOUS and (b) HUMORAL factors influence its expulsion:-

(a) *Sight, Smell, Taste of Food.* } reflex → *Contraction of Gall Bladder with Relaxation of Sphincter of Oddi.* } BILE enters Duodenum.

(b) *Fat in Duodenum* → *Secretion of Hormone CHOLE-CYSTOKININ.*

FUNCTIONS of BILE

BILE SALTS help to EMULSIFY fat, ACTIVATE pancreatic lipase and promote ABSORPTION of fat.

most reabsorbed → LIVER where they stimulate further excretion of Bile (*CHOLERETIC ACTION*)

BILE PIGMENTS *some reabsorbed* → LIVER
(*from breakdown of Haemoglobin in Liver*)
↓ Re-excretion in Bile.
General Circulation
↓
KIDNEY
UROBILINOGEN
↓
UROBILIN of URINE
↓

changed by action of bacteria

STERCOBILIN
(Brown pigment of FAECES)

Other substances, e.g. certain drugs, can be taken out of the blood stream by the Liver Cells and excreted in the Bile.

36

SMALL INTESTINE

The Small Intestine is a long muscular tube-over 21 feet in length.
It receives CHYME in small quantities from the STOMACH; PANCREATIC
JUICE from the PANCREAS; BILE from the GALL BLADDER.

DUODENUM (10"-12")

It has 4 Coats:-
1. Outer <u>SEROUS COAT</u> of Peritoneum *(with vessels & nerves).*

2. <u>MUSCULAR COAT</u>
Smooth muscle - 2 layers:-
Outer- *LONGITUDINAL*
Inner- *CIRCULAR*

STOMACH

3. <u>SUBMUCOUS COAT</u> with
fibrous tissue, bv's.
and (in duodenum only)
Brunner's (mucous) glands

JEJUNUM (8')

4. <u>MUCOUS COAT</u>
(or lining)

×10

Gradual
transition
to

ILEUM (12')

×10

VILLI-
*finger-like projections with
STRIATED BORDER EPITHELIUM
-increase surface area for*
ABSORPTION

ILEO-
CAECAL
VALVE

20 to 30
*Aggregations of
Lymph follicles in*
PEYER'S PATCH *form part of
the ileum's defence mechanism
against bacteria.*

×10

<u>MOVEMENTS</u> of the wall mix
food with digestive juices,
promote absorption and move
the residue along the tube.

Between the muscular layers
lies AUERBACH'S Nerve Plexus
through which peristaltic
movements are controlled.

<u>CRYPTS of LIEBERKÜHN</u>
secrete alkaline Intestinal
Juice (Succus Entericus).
PANETH CELLS at the base of
Crypts secrete ENZYMES of
SUCCUS ENTERICUS.
EREPSIN-Splits peptides to
amino acids.
ENTEROKINASE activates
Pancreatic Trypsinogen.
AMYLASE-Splits Starch to Maltose.
MALTASE⎤ Complete digestion of
LACTASE⎬ carbohydrates to
SUCRASE⎦   simple sugars.
LIPASE-Splits fats to lower
glycerides, fatty acids
and glycerol.

The MUCOSA also secretes
HORMONES—
e.g. ENTEROGASTRONE,
SECRETIN,
PANCREOZYMIN,
CHOLECYSTOKININ —
to help regulate flow of Gastro-
intestinal secretions.
*(SECRETIN is one of the factors
regulating flow of Succus
Entericus.)*

MOVEMENTS of the SMALL INTESTINE

The Duodenum receives food in small quantities from the Stomach.
The mixture of food and digestive juices — Chyme – is passed along the length of the Small Intestine.

TWO TYPES of MOVEMENT

SEGMENTATION —
*Rhythmical alternating
contractions and relaxations*

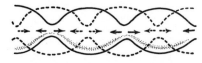

— These "shuttling" movements serve to mix CHYME and to bring it into contact with the absorptive mucosa, i.e. DIGESTION and ABSORPTION are promoted.

This type of movement is MYOGENIC, i.e. it is the property of the smooth muscle cells. It does not depend on a nervous mechanism.

PERISTALSIS —
Food probably acts as stimulus to stretch receptors in muscle and perhaps in mucous membrane

Circular muscle   Muscle in front
behind bolus *CONTRACTS*   /of bolus *RELAXES*

— Waves of this contraction move the food along the canal.

The contraction behind the bolus sweeps it into the relaxed portion of the tube ahead.

*This type of movement is NEUROGENIC, i.e. it is carried out reflexly through the nerve plexuses in the wall of the
tube.*

EMPTYING

ILEO-CAECAL VALVE ------- opens and closes during
Formed by thickening of digestion to allow spurts of
circular muscle layer fluid material from the Ileum
to enter the Large Intestine.

ILEUM

CAECUM

This reflex is initiated when food enters the STOMACH. Motor nerve messages are carried to the wall of the tube by the autonomic nerves. (see page 149)

Meals of different composition travel along the intestine at different rates. DIGESTION and ABSORPTION of food are usually complete by the time the residue reaches the Ileo-caecal valve.

38

ABSORPTION in SMALL INTESTINE

Absorption of most digested foodstuffs occurs in the Small Intestine through the Striated Border Epithelium covering the Villi. *(see page 6)*

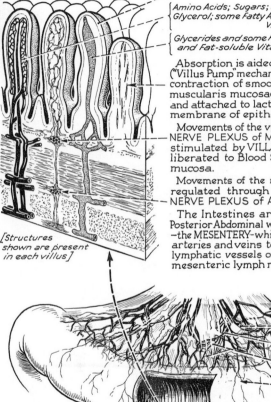

Amino Acids; Sugars; Minerals; Glycerol; some Fatty Acids and Vitamins — into CAPILLARIES

Glycerides and some Fatty Acids and Fat-soluble Vitamins — into LACTEALS

Absorption is aided by movements of villi ("Villus Pump" mechanism) brought about by contraction of smooth muscle – extensions of muscularis mucosae – present in core of villus and attached to lacteal and to basement membrane of epithelium.

Movements of the villi are regulated through NERVE PLEXUS of MEISSNER and perhaps stimulated by VILLIKININ – a hormone liberated to Blood Stream by Duodenal mucosa.

Movements of the muscular coats are regulated through the NERVE PLEXUS of AUERBACH.

The Intestines are suspended from the Posterior Abdominal wall by a delicate membrane – the MESENTERY – which carries the mesenteric arteries and veins to and from the gut; and lymphatic vessels on their way via the mesenteric lymph nodes to Thoracic Duct.

[Structures shown are present in each villus]

——————Mesentery

———— Branches of Mesenteric Vein

——— Branches of Mesenteric Artery

———— Lymphatic Vessels

——— Nerves

LARGE INTESTINE

(Total length over 6')

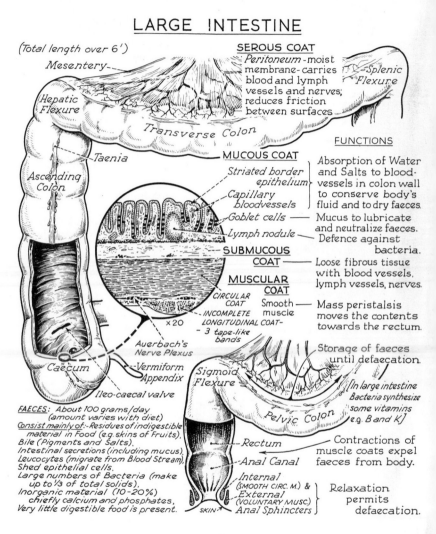

Mesentery

Hepatic Flexure

Transverse Colon

Splenic Flexure

SEROUS COAT

Peritoneum - moist membrane - carries blood and lymph vessels and nerves; reduces friction between surfaces

Taenia

Ascending Colon

MUCOUS COAT

Striated border epithelium

Capillary bloodvessels

Goblet cells

Lymph nodule

SUBMUCOUS COAT

MUSCULAR COAT

CIRCULAR COAT

INCOMPLETE LONGITUDINAL COAT - 3 tape-like bands

×20

Smooth muscle

Auerbach's Nerve Plexus

Vermiform Appendix

Caecum

Ileo-caecal valve

Sigmoid Flexure

Pelvic Colon

Rectum

Anal Canal

Internal (SMOOTH CIRC. M.) & External (VOLUNTARY MUSC.) Anal Sphincters

SKIN

FUNCTIONS

Absorption of Water and Salts to bloodvessels in colon wall to conserve body's fluid and to dry faeces.

Mucus to lubricate and neutralize faeces. Defence against bacteria.

Loose fibrous tissue with blood vessels, lymph vessels, nerves.

Mass peristalsis moves the contents towards the rectum.

Storage of faeces until defaecation.

[In large intestine Bacteria synthesize some vitamins e.g. B and K]

Contractions of muscle coats expel faeces from body.

Relaxation permits defaecation.

FAECES: About 100 grams/day (amount varies with diet)
Consist mainly of:- Residues of indigestible material in Food (e.g. skins of Fruits),
Bile (Pigments and Salts),
Intestinal secretions (including mucus),
Leucocytes (migrate from Blood Stream),
Shed epithelial cells,
Large numbers of Bacteria (make up to 1/3 of total solids),
Inorganic material (10-20%) chiefly calcium and phosphates,
Very little digestible food is present.

MOVEMENTS of the LARGE INTESTINE

The Ileo-caecal Valve opens and closes during digestion. Peristaltic waves sweep semi-fluid contents of Ileum through the relaxed valve. During its stay in the Large Intestine faecal matter is subjected to –

TWO TYPES OF MOVEMENT

"TONE WAVES" – – – – – – – – – – – – – – – → Ensure prolonged contact
run forwards –
and backwards from
Cannon's Point to the
Ileum – (sometimes
mistakenly called
"Anti-peristalsis")
These waves "die out"

CANNON'S POINT

of contents with mucosa
—and promote absorption
of Water and Salt from
faeces—
This type of movement is
MYOGENIC

MASS PERISTALSIS
Strong waves – – – – – – – – – – – – – – – → Empty Transverse Colon
at infrequent
intervals start
at upper end of
Ascending Colon

and sweep faeces into the
Descending and Pelvic
Colons and into Rectum—
—*NEUROGENIC*

[Reflex often initiated by
passage of food into Stomach
(Gastro-colic Reflex)]

EMPTYING
(DEFAECATION)
Complex
REFLEX act

Stimulus
Passage of
Faeces into
the Rectum
distends
wall

[
+
passage
through
anal
canal
]

SYMPATHETIC
NERVES

(+)

FAECES

PELVIC
NERVES
PARASYMP.
NERVES

(−)

PUDENDAL
NERVES

(−)

SPINAL CORD

Sensations rise to level of consciousness →
Voluntary decision → Impulses to
inhibit or permit reflex evacuation

Outgoing nerve messages →
Powerful peristaltic
contractions of Descending
Colon, Pelvic Colon and Rectum.

[*Preceded by inspiratory*
descent of diaphragm and
voluntary contraction of
abdominal muscles to raise
intra-abdominal pressure]

Contraction of Pelvic Floor
muscles with Relaxation of
Anal Sphincter
↓
Evacuation of Faeces

NERVOUS CONTROL of GUT MOVEMENTS

The movements of the Alimentary Canal are carried out automatically and, for the most part, beneath the level of consciousness.

Movements in the wall of the Alimentary Canal are
either
- *MYOGENIC — a property of the smooth muscle.*
- *NEUROGENIC — dependent on the Intrinsic Nerve Plexuses.*

They can occur even after *Extrinsic* nerves to the tract have been cut. Normally, however, impulses travelling in these nerves of the SYMPATHETIC and PARASYMPATHETIC Systems, from the CONTROLLING CENTRES of the AUTONOMIC NERVOUS SYSTEM in the BRAIN, *influence* and *co-ordinate* events in the whole tract.

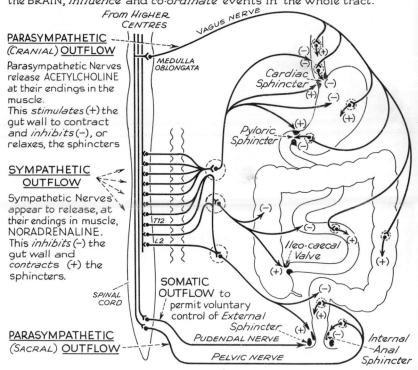

PARASYMPATHETIC (*CRANIAL*) OUTFLOW

Parasympathetic Nerves release ACETYLCHOLINE at their endings in the muscle.
This *stimulates* (+) the gut wall to contract and *inhibits* (−), or relaxes, the sphincters

SYMPATHETIC OUTFLOW

Sympathetic Nerves appear to release, at their endings in muscle, NORADRENALINE.
This *inhibits* (−) the gut wall and *contracts* (+) the sphincters.

PARASYMPATHETIC (*SACRAL*) OUTFLOW

From HIGHER CENTRES
VAGUS NERVE
MEDULLA OBLONGATA
Cardiac Sphincter
Pyloric Sphincter
Ileo-caecal Valve
SPINAL CORD
T12
L2

SOMATIC OUTFLOW to permit voluntary control of *External Sphincter*
PUDENDAL NERVE
PELVIC NERVE
Internal Anal Sphincter

TRANSPORT and UTILIZATION of FOODSTUFFS

After absorption the NUTRIENTS are transported in :-

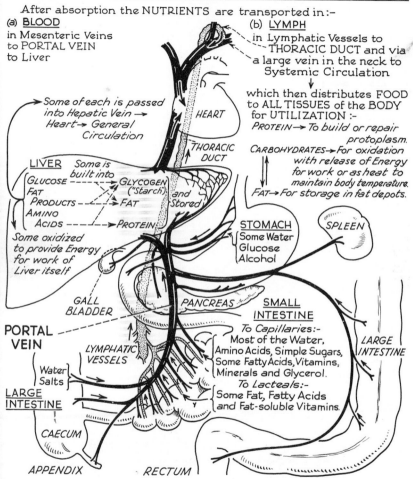

(a) BLOOD
in Mesenteric Veins
to PORTAL VEIN
to Liver

(b) LYMPH
in Lymphatic Vessels to
THORACIC DUCT and via
a large vein in the neck to
Systemic Circulation

which then distributes FOOD
to ALL TISSUES of the BODY
for UTILIZATION :-

PROTEIN → To build or repair
protoplasm.

CARBOHYDRATES → For oxidation
with release of Energy
for work or as heat to
maintain body temperature.

FAT → For storage in fat depots.

Some of each is passed
into Hepatic Vein →
Heart → General
Circulation

HEART

THORACIC
DUCT

LIVER Some is
built into

GLUCOSE - - - - → GLYCOGEN
("Starch")
and
Stored

FAT
PRODUCTS - - - → FAT

AMINO
ACIDS - - - - → PROTEIN

Some oxidized
to provide Energy
for work of
Liver itself

STOMACH
Some Water
Glucose
Alcohol

SPLEEN

GALL
BLADDER

PANCREAS

**SMALL
INTESTINE**

To Capillaries :-
Most of the Water,
Amino Acids, Simple Sugars,
Some Fatty Acids, Vitamins,
Minerals and Glycerol.

To Lacteals :-
Some Fat, Fatty Acids
and Fat-soluble Vitamins.

**PORTAL
VEIN**

LYMPHATIC
VESSELS

**LARGE
INTESTINE**

Water
Salts

**LARGE
INTESTINE**

CAECUM

APPENDIX

RECTUM

43

HEAT BALANCE

ENERGY is released in cells by OXIDATION. It is used for work and to keep body warm. The Body Temperature is kept relatively constant *(with a slight fluctuation throughout the 24 hours)* in spite of wide variations in Environmental Temperature and Heat Production.

HEAT PRODUCTION ——— *must balance* ——— HEAT LOSS

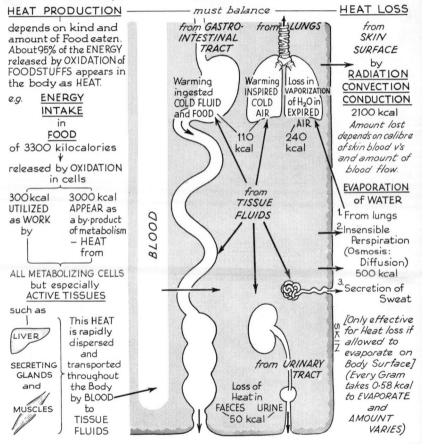

HEAT PRODUCTION

depends on kind and amount of Food eaten. About 95% of the ENERGY released by OXIDATION of FOODSTUFFS appears in the body as HEAT.

e.g.   **ENERGY INTAKE**
in
FOOD
of 3300 kilocalories
↓
released by OXIDATION in cells

| 300 kcal UTILIZED as WORK by | 3000 kcal APPEAR as a by-product of metabolism – HEAT from |

ALL METABOLIZING CELLS
but especially
ACTIVE TISSUES

such as

LIVER

SECRETING GLANDS and

MUSCLES

This HEAT is rapidly dispersed and transported throughout the Body by BLOOD to TISSUE FLUIDS

from GASTRO-INTESTINAL TRACT

Warming ingested COLD FLUID and FOOD

110 kcal

from LUNGS

Warming INSPIRED COLD AIR

Loss in VAPORIZATION of H_2O in EXPIRED AIR

240 kcal

from TISSUE FLUIDS

BLOOD

from URINARY TRACT

Loss of Heat in FAECES   URINE
50 kcal

HEAT LOSS

from SKIN SURFACE
by
RADIATION CONVECTION CONDUCTION
2100 kcal
Amount lost depends on calibre of skin blood v's and amount of blood flow.

EVAPORATION of WATER

1. From lungs
2. Insensible Perspiration (Osmosis: Diffusion) 500 kcal
3. Secretion of Sweat

SKIN

[Only effective for Heat loss if allowed to evaporate on Body Surface] (Every Gram takes 0·58 kcal to EVAPORATE and AMOUNT VARIES)

44

MAINTENANCE of BODY TEMPERATURE

Any tendency for the BODY TEMPERATURE to *RISE*
as by

1. INCREASED HEAT PRODUCTION — is balanced by **INCREASED HEAT LOSS**

by increased cellular
<u>OXIDATION of FOODSTUFFS</u>
as occurs e.g. with
<u>MUSCULAR ACTIVITY</u>

<u>EXTRA HEAT</u>
is dispersed
quickly by Blood Stream

Rise in Blood
Temperature
affects
HYPOTHALAMUS

2. HOT ENVIRONMENTAL TEMPERATURE
i.e. above BODY
TEMPERATURE (37°C)

BLOOD
VESSELS
DILATED

FAT

Reduced
SYMPATHETIC
VASOCONSTRICTOR
tone to
SECRETOMOTOR

HEAT SENSITIVE
nerve endings
in SKIN are stimulated

INGOING
NERVE
IMPULSES

OUTGOING
NERVE IMPULSES

1. SKIN BLOOD VESSELS DILATE

*more blood to skin
surface→ increased heat
loss from skin by –*
RADIATION ⎱ (Cannot
CONVECTION ⎰ occur if air
CONDUCTION ⎰ temperature
is above
body's)
*Increased by
voluntary ingestion of
cold foods and fluids
and by use of fans.*

2. SWEAT GLANDS SECRETE
*Increased heat loss by
evaporation from skin
surface (unless atmosphere
is already water saturated
as e.g. in Tropics)*

3. DIMINISHED HEAT INSULATION
*by voluntary reduction
of CLOTHING worn*

4. DIMINISHED HEAT PRODUCTION
SKELETAL MUSCLE *'tone'*
*reduced – and often
voluntary relaxation
→ less work done →
less heat produced*

5. REDUCTION of 'ENERGY INTAKE'
by voluntary restriction of
<u>*PROTEIN in DIET*</u>

*[The increase in activity during the day
probably accounts for the gradual Physiological
rise in body temperature from about 96·5°F
(35·8°C) in the early morning to about 99·2°F
(37·3°C) in the late afternoon.*

*Unless exercise is very strenuous or environ-
ment is very hot and humid these measures* → **RESTORE BODY TEMPERATURE**
to NORMAL

4

MAINTENANCE of BODY TEMPERATURE

Any tendency for the BODY TEMPERATURE to *FALL* as by

1. DIMINISHED HEAT PRODUCTION — is balanced by **DIMINISHED HEAT LOSS**

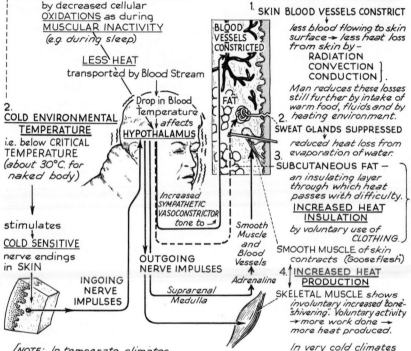

by decreased cellular <u>OXIDATIONS</u> as during <u>MUSCULAR INACTIVITY</u> (*e.g. during sleep*)

<u>LESS HEAT</u> transported by Blood Stream

Drop in Blood Temperature affects **HYPOTHALAMUS**

2. COLD ENVIRONMENTAL TEMPERATURE i.e. below CRITICAL TEMPERATURE (*about 30°C. for naked body*)

stimulates <u>COLD SENSITIVE</u> nerve endings in SKIN

INGOING NERVE IMPULSES

Increased SYMPATHETIC VASOCONSTRICTOR *tone to* →

OUTGOING NERVE IMPULSES

Suprarenal Medulla

Adrenaline

BLOOD VESSELS CONSTRICTED

FAT

Smooth Muscle and Blood Vessels

1. SKIN BLOOD VESSELS CONSTRICT

less blood flowing to skin surface → *less heat loss from skin by* –
RADIATION
CONVECTION
CONDUCTION.
Man reduces these losses still further by intake of warm food, fluids and by heating environment.

2. SWEAT GLANDS SUPPRESSED
reduced heat loss from evaporation of water.

3. SUBCUTANEOUS FAT –
an insulating layer through which heat passes with difficulty.
INCREASED HEAT INSULATION
by voluntary use of CLOTHING.
SMOOTH MUSCLE *of skin contracts* (Gooseflesh)

4. INCREASED HEAT PRODUCTION
SKELETAL MUSCLE *shows involuntary increased tone "shivering". Voluntary activity* → *more work done* → *more heat produced.*

In very cold climates there is a tendency to

5. INCREASE 'ENERGY INTAKE'
by voluntary increase in <u>DIETARY PROTEIN</u> *which has stimulating effect on metabolism.*

[NOTE: In temperate climates environmental temperature is usually lower than body temperature so that there is a continuous loss of heat from body surface.]

Unless environmental temperature is very low these measures tend to — <u>RESTORE BODY TEMPERATURE</u> to NORMAL

CHAPTER 3.
TRANSPORT SYSTEM

All cells are bathed by TISSUE FLUID. It is from solution in this fluid that O_2 and food materials diffuse into each cell; to it waste products including CO_2 diffuse out of the cells.

The CARDIOVASCULAR SYSTEM is the TRANSPORT SYSTEM which conveys these materials to and from the tissues.

This simplified diagram gives a concept of the general plan of the circulation.

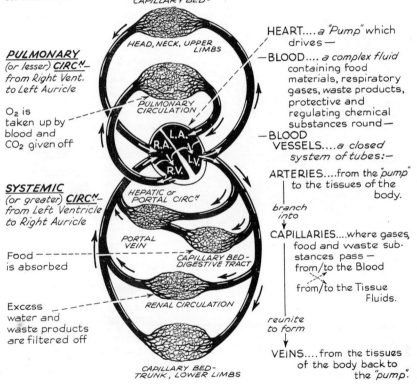

CAPILLARY BED -

HEAD, NECK, UPPER LIMBS

PULMONARY
(or lesser) **CIRC^N**
from Right Vent.
to Left Auricle

O_2 is taken up by blood and CO_2 given off

PULMONARY CIRCULATION

R.A. L.A.

R.V. L.V.

SYSTEMIC
(or greater) **CIRC^N**
from Left Ventricle
to Right Auricle

HEPATIC or PORTAL CIRC^N

Food is absorbed

PORTAL VEIN

CAPILLARY BED - DIGESTIVE TRACT

Excess water and waste products are filtered off

RENAL CIRCULATION

CAPILLARY BED - TRUNK, LOWER LIMBS

HEART....a "Pump" which drives —

—BLOOD.... a complex fluid containing food materials, respiratory gases, waste products, protective and regulating chemical substances round—

—BLOOD VESSELS....a closed system of tubes:-

ARTERIES....from the "pump" to the tissues of the body.

branch into

CAPILLARIES....where gases, food and waste substances pass — from/to the Blood

from/to the Tissue Fluids.

reunite to form

VEINS....from the tissues of the body back to the "pump".

HEART

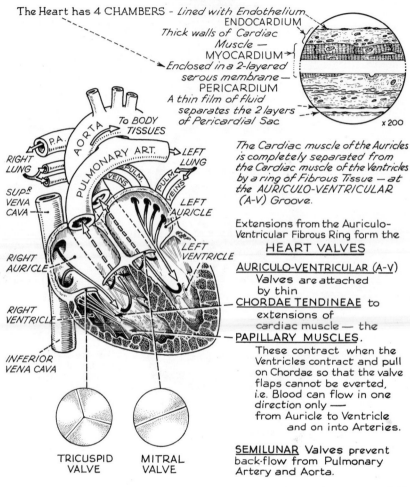

The Heart has 4 CHAMBERS - *Lined with Endothelium,*
ENDOCARDIUM
Thick walls of Cardiac
Muscle —
MYOCARDIUM
Enclosed in a 2-layered
serous membrane —
PERICARDIUM
A thin film of fluid
separates the 2 layers
of Pericardial Sac

x 200

P.A.

AORTA   To BODY
            TISSUES

RIGHT
LUNG   PULMONARY ART.   LEFT
                                LUNG

PULM VEINS   PULM VEINS

SUPᴿ
VENA
CAVA   LEFT
                                AURICLE

RIGHT
AURICLE   LEFT
                                VENTRICLE

RIGHT
VENTRICLE

INFERIOR
VENA CAVA

*The Cardiac muscle of the Auricles
is completely separated from
the Cardiac muscle of the Ventricles
by a ring of Fibrous Tissue — at
the AURICULO-VENTRICULAR
(A-V) Groove.*

Extensions from the Auriculo-
Ventricular Fibrous Ring form the
HEART VALVES

<u>AURICULO-VENTRICULAR (A-V)</u>
Valves are attached
by thin
<u>CHORDAE TENDINEAE</u> to
extensions of
cardiac muscle — the
<u>PAPILLARY MUSCLES</u>.
These contract when the
Ventricles contract and pull
on Chordae so that the valve
flaps cannot be everted,
i.e. Blood can flow in one
direction only —
from Auricle to Ventricle
and on into Arteries.

<u>SEMILUNAR</u> Valves prevent
back-flow from Pulmonary
Artery and Aorta.

TRICUSPID   MITRAL
VALVE   VALVE

HEART

The human heart is really a *DOUBLE PUMP* — each quite
separate from the other

RIGHT   *LEFT*

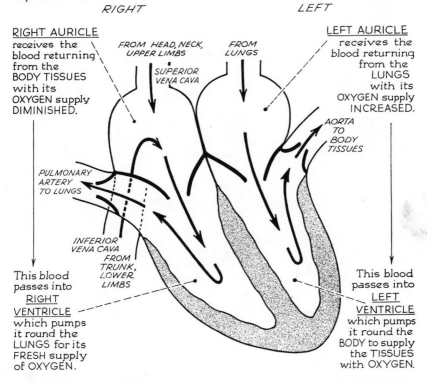

<u>RIGHT AURICLE</u>
receives the
blood returning
from the
BODY TISSUES
with its
OXYGEN supply
DIMINISHED.

FROM HEAD, NECK,
UPPER LIMBS

SUPERIOR
VENA CAVA

FROM
LUNGS

<u>LEFT AURICLE</u>
receives the
blood returning
from the
LUNGS
with its
OXYGEN supply
INCREASED.

AORTA
TO
BODY
TISSUES

PULMONARY
ARTERY
TO LUNGS

INFERIOR
VENA CAVA
FROM
TRUNK,
LOWER
LIMBS

This blood
passes into
<u>RIGHT</u>
<u>VENTRICLE</u>
which pumps
it round the
LUNGS for its
FRESH supply
of OXYGEN.

This blood
passes into
<u>LEFT</u>
<u>VENTRICLE</u>
which pumps
it round the
BODY to supply
the TISSUES
with OXYGEN.

This diagram simplifies the structure of the heart to make it
easier to understand the function of its various parts.

CARDIAC CYCLE

Diagrammatic representation of the sequence of events in the heart during ONE heart beat.

DIASTOLE *[Period of Relaxation – i.e. when heart is resting]*

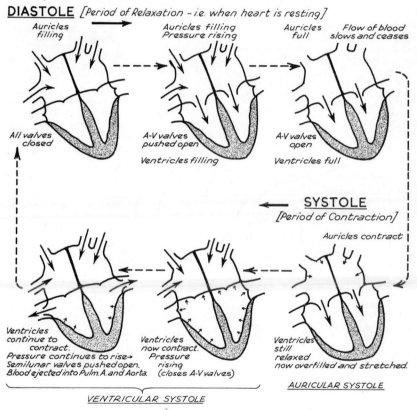

Auricles filling

Auricles filling
Pressure rising

Auricles full

Flow of blood slows and ceases

All valves closed

A-V valves pushed open

Ventricles filling

A-V valves open

Ventricles full

SYSTOLE
[Period of Contraction]

Auricles contract

Ventricles continue to contract.
Pressure continues to rise→
Semilunar valves pushed open.
Blood ejected into Pulm. A. and Aorta.

Ventricles now contract.
Pressure rising
(closes A-V valves)

Ventricles still relaxed
now overfilled and stretched.

VENTRICULAR SYSTOLE

AURICULAR SYSTOLE

The total cycle of events takes about 0·8 second when heart is beating 75 times per minute.

HEART SOUNDS

During each CARDIAC CYCLE 2 HEART SOUNDS can be heard through a STETHOSCOPE applied to the CHEST WALL.

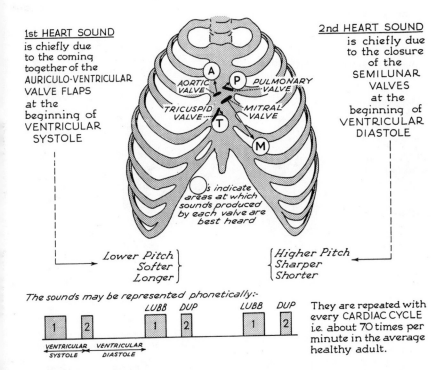

1st HEART SOUND
is chiefly due to the coming together of the AURICULO-VENTRICULAR VALVE FLAPS at the beginning of VENTRICULAR SYSTOLE

2nd HEART SOUND
is chiefly due to the closure of the SEMILUNAR VALVES at the beginning of VENTRICULAR DIASTOLE

AORTIC VALVE — A

P — PULMONARY VALVE

TRICUSPID VALVE — T

MITRAL VALVE — M

○s indicate areas at which sounds produced by each valve are best heard

Lower Pitch
Softer
Longer

Higher Pitch
Sharper
Shorter

The sounds may be represented phonetically:-

LUBB   DUP   LUBB   DUP
1   2   1   2   1   2

VENTRICULAR SYSTOLE   VENTRICULAR DIASTOLE

They are repeated with every CARDIAC CYCLE i.e. about 70 times per minute in the average healthy adult.

If the valves have been damaged by disease additional murmurs can be heard as the blood flows forwards through narrowed valves or leaks backwards through incompetent valves.

Occasionally additional sounds are heard over normal healthy hearts.

ORIGIN and CONDUCTION of HEART BEAT.

The rhythmic contraction of the heart is called the HEART BEAT. The impulse to contract is generated in specialized NODAL TISSUE in the wall of the RIGHT AURICLE.

Impulses are discharged rhythmically from this SINO-AURICULAR NODE *(The "Pacemaker")*

The wave of excitation spreads throughout the muscle of both AURICLES

which are then excited to contract

The Right Auricle starts contracting just before Left Auricle.

The impulse is picked up by another mass of NODAL TISSUE — the AURICULO-VENTRICULAR NODE — and relayed by PURKINJE TISSUE — *(in Bundle of His and its branches)* lying beneath endocardium on the interventricular septum.

This relays the impulse to contract to the muscle of both VENTRICLES .... which then contract together

A ring of fibrous tissue separates Auricles from Ventricles. The heart beat is not transmitted from Auricles to Ventricles directly by ordinary CARDIAC muscle.

x 150

The wave of excitation spreading through heart muscle is accompanied by electrical changes. These can be recorded by an electrocardiograph.

In complete heart block the impulse is not transmitted to the Ventricles and they contract independently. In incomplete heart block every 2nd. or 3rd. impulse gets through.

Although the heart initiates its own impulse to contract its activity is finely adjusted, to meet the body's constantly changing needs, by nervous impulses discharged from Controlling Centres in the Brain and Spinal Cord. The Sympathetic nerves increase the rate and force of the heart beat: the Parasympathetic nerves slow heart and reduce force of contraction *(see pp. 149,150)*.

BLOOD VESSELS

The system of tubes through which the heart pumps blood.

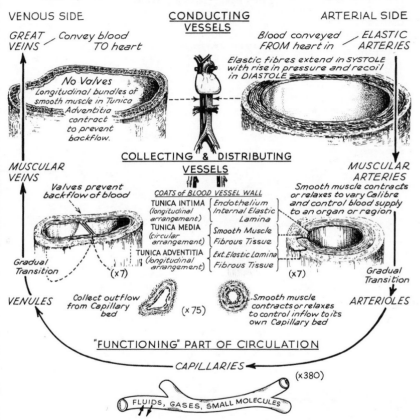

VENOUS SIDE

CONDUCTING VESSELS

ARTERIAL SIDE

GREAT VEINS — *Convey blood TO heart*

Blood conveyed FROM heart in — *ELASTIC ARTERIES*

Elastic fibres extend in SYSTEM with rise in pressure and recoil in DIASTOLE

No Valves
Longitudinal bundles of smooth muscle in Tunica Adventitia contract to prevent backflow.

COLLECTING & DISTRIBUTING VESSELS

MUSCULAR VEINS

MUSCULAR ARTERIES

Valves prevent backflow of blood

Smooth muscle contracts or relaxes to vary Calibre and control blood supply to an organ or region

COATS of BLOOD VESSEL WALL

TUNICA INTIMA (longitudinal arrangement)	Endothelium / Internal Elastic Lamina
TUNICA MEDIA (circular arrangement)	Smooth Muscle / Fibrous Tissue
TUNICA ADVENTITIA (longitudinal arrangement)	Ext. Elastic Lamina / Fibrous Tissue

Gradual Transition

(x7)

(x7)

Gradual Transition

VENULES

Collect outflow from Capillary bed

(x 75)

Smooth muscle contracts or relaxes to control inflow to its own Capillary bed

ARTERIOLES

"FUNCTIONING" PART OF CIRCULATION

CAPILLARIES

(x380)

FLUIDS, GASES, SMALL MOLECULES

Only from CAPILLARIES can Blood give up food and oxygen to tissues; and receive waste products and carbon dioxide from tissues.

BLOOD PRESSURE

Each Ventricle at each heart beat ejects forcibly about 70c.c. of Blood into the Blood Vessels. All of this blood cannot pass through arterioles into capillaries and veins before the onset of the next heart beat. This means that roughly 5/8 of the CARDIAC OUTPUT at each heart beat has to be stored during systole and passed on during diastole.

CONDUCTING ARTERIES are always more or less stretched.

PERIPHERAL RESISTANCE is offered to the passage of blood from arterial to venous side of the system chiefly by partial constriction ('Tone') of smooth muscle in walls of Arterioles. (*The calibre is regulated by action of Parasympathetic and Sympathetic Nervous System* – [see pages 149-150]).

These factors are largely responsible for the considerable pressure of the blood in the Arterial System. The pressure is highest at the height of the heart's contraction, i.e. SYSTOLIC BLOOD PRESSURE, and lowest when the heart is relaxing, i.e. DIASTOLIC BLOOD PRESSURE.

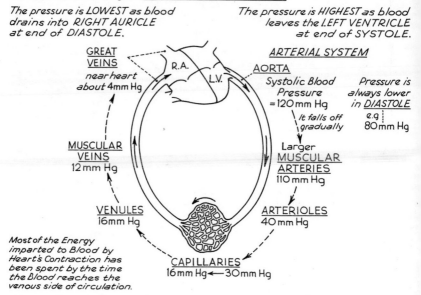

The pressure is LOWEST as blood drains into RIGHT AURICLE at end of DIASTOLE.

The pressure is HIGHEST as blood leaves the LEFT VENTRICLE at end of SYSTOLE.

GREAT VEINS
near heart
about 4mm Hg

R.A. L.V.

ARTERIAL SYSTEM
AORTA
Systolic Blood Pressure = 120 mm Hg

Pressure is always lower in DIASTOLE
e.g. 80 mm Hg

It falls off gradually

MUSCULAR VEINS
12 mm Hg

Larger MUSCULAR ARTERIES
110 mm Hg

VENULES
16mm Hg

ARTERIOLES
40 mm Hg

Most of the Energy imparted to Blood by Heart's Contraction has been spent by the time the Blood reaches the venous side of circulation.

CAPILLARIES
16mm Hg ← 30mm Hg

NOTE :- Any alteration in the TOTAL AMOUNT *or* VISCOSITY *of Blood will also affect BLOOD PRESSURE.*

MEASUREMENT of ARTERIAL BLOOD PRESSURE

The Arterial Blood Pressure is measured in man by means of a
SPHYGMOMANOMETER.

This consists of a
RUBBER BAG _(covered
with a cloth envelope)_
which is wrapped
round the UPPER ARM
over the BRACHIAL
ARTERY.

One tube connects the
inside of the bag with
a MANOMETER
containing MERCURY.

Another tube connects
the inside of the bag to
a hand operated PUMP
with a release VALVE.

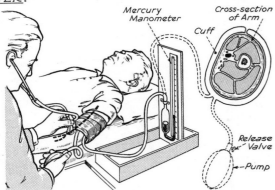

Mercury Manometer   Cross-section of Arm

Cuff

Release Valve

Pump

METHOD

_Air is pumped into the rubber bag till
pressure in cuff is greater than pressure
in artery even during heart's systole —
    — Artery is then closed down during_ SYSTOLE   and   DIASTOLE
_[At same time air is pushing up mercury
           column in manometer.]_

_By releasing valve on pump the pressure
in cuff is gradually reduced till maximum
pressure in artery just overcomes pressure in
cuff—Some blood begins to spurt through during_ SYSTOLE   — _Artery still closed
                                           during DIASTOLE_

_At this point FAINT rhythmical TAPPING SOUNDS
begin to be heard through STETHOSCOPE. The
height of mercury in millimetres is taken as the
SYSTOLIC Blood Pressure (e.g. 120 mm Hg)_

_Pressure in cuff is reduced still further
till it is just less than the lowest pressure
in artery towards the end of diastole
(i.e. just before next heart beat) —_

_   — Blood flow is unimpeded during_ SYSTOLE _and_ DIASTOLE
_The sounds stop. The height of mercury in the manometer at this point
is taken as the DIASTOLIC Blood Pressure (e.g. about 80 mm Hg)_

These values differ with SEX, AGE, EXERCISE, SLEEP, etc.

ELASTIC ARTERIES

The large <u>CONDUCTING ARTERIES</u> near the <u>HEART</u> are
<u>ELASTIC ARTERIES</u>

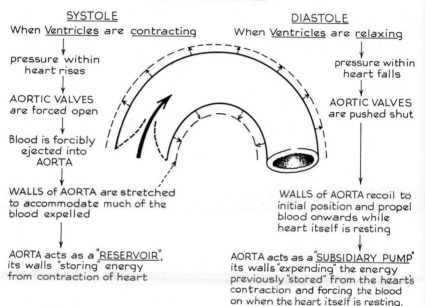

<u>SYSTOLE</u>
When <u>Ventricles</u> are <u>contracting</u>

↓

pressure within
heart rises

↓

AORTIC VALVES
are forced open

↓

Blood is forcibly
ejected into
AORTA

↓

WALLS of AORTA are stretched
to accommodate much of the
blood expelled

↓

AORTA acts as a <u>"RESERVOIR"</u>,
its walls "storing" energy
from contraction of heart

<u>DIASTOLE</u>
When <u>Ventricles</u> are <u>relaxing</u>

↓

pressure within
heart falls

↓

AORTIC VALVES
are pushed shut

↓

WALLS of AORTA recoil to
initial position and propel
blood onwards while
heart itself is resting

↓

AORTA acts as a <u>"SUBSIDIARY PUMP"</u>
its walls "expending" the energy
previously "stored" from the heart's
contraction and forcing the blood
on when the heart itself is resting.

As blood is pumped from the heart during systole, this distension
and increase in pressure which starts in the aorta passes along the
whole arterial system as a wave — the *pulse wave*.

The expansion and subsequent relaxation of the wall of the radial
artery can be felt as "The PULSE" at the wrist.

*A great increase in Blood Pressure can result if
these walls lose some of their elasticity with age or disease
and can no longer stretch readily to accommodate so much of
the heart's output during systole.*

CAPILLARIES : EXCHANGE of WATER & SOLUTES

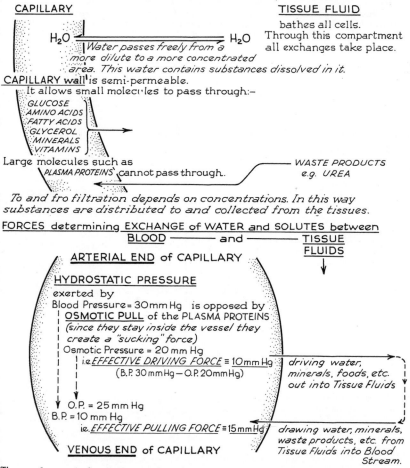

CAPILLARY

TISSUE FLUID

bathes all cells.
Through this compartment
all exchanges take place.

H_2O — *Water passes freely from a more dilute to a more concentrated area. This water contains substances dissolved in it.* — H_2O

CAPILLARY wall is semi-permeable.
It allows small molecules to pass through:-

GLUCOSE
AMINO ACIDS
FATTY ACIDS
GLYCEROL
MINERALS
VITAMINS

Large molecules such as
PLASMA PROTEINS cannot pass through.

WASTE PRODUCTS
e.g. UREA

To and fro filtration depends on concentrations. In this way substances are distributed to and collected from the tissues.

FORCES determining EXCHANGE of WATER and SOLUTES between
BLOOD —— and —— TISSUE FLUIDS

ARTERIAL END of CAPILLARY

HYDROSTATIC PRESSURE
exerted by
Blood Pressure = 30 mm Hg is opposed by
OSMOTIC PULL of the PLASMA PROTEINS
(since they stay inside the vessel they create a "sucking" force)
Osmotic Pressure = 20 mm Hg
i.e. *EFFECTIVE DRIVING FORCE* ≡ 10 mm Hg
(B.P. 30 mm Hg − O.P. 20 mm Hg)

driving water, minerals, foods, etc. out into Tissue Fluids

O.P. = 25 mm Hg
B.P. = 10 mm Hg
i.e. *EFFECTIVE PULLING FORCE* ≡ 15 mm Hg

drawing water, minerals, waste products, etc. from Tissue Fluids into Blood Stream.

VENOUS END of CAPILLARY

These exchanges in Capillaries result in a continuous "turn-over" and renewal of Tissue Fluids.

VEINS: VENOUS RETURN

Capillaries unite to form *Veins* which *convey blood back to the heart.* By the time blood reaches the veins much of the force imparted to it by the heart's contraction has been spent.

<u>VENOUS RETURN</u> to the heart depends on various factors:-

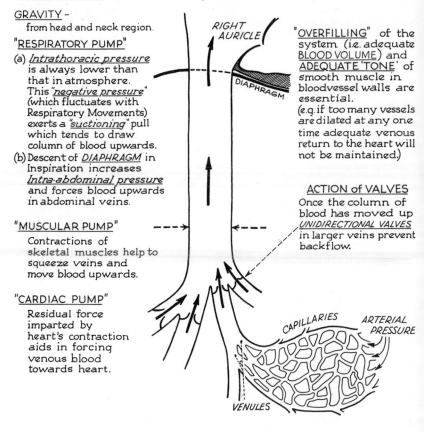

GRAVITY -
from head and neck region.

"<u>RESPIRATORY PUMP</u>"

(a) *Intrathoracic pressure* is always lower than that in atmosphere. This *"negative pressure"* (which fluctuates with Respiratory Movements) exerts a *"suctioning"* pull which tends to draw column of blood upwards.

(b) Descent of *DIAPHRAGM* in Inspiration increases *Intra-abdominal pressure* and forces blood upwards in abdominal veins.

"<u>MUSCULAR PUMP</u>"

Contractions of skeletal muscles help to squeeze veins and move blood upwards.

"<u>CARDIAC PUMP</u>"

Residual force imparted by heart's contraction aids in forcing venous blood towards heart.

RIGHT AURICLE

DIAPHRAGM

"<u>OVERFILLING</u>" of the system (i.e. adequate <u>BLOOD VOLUME</u>) and <u>ADEQUATE 'TONE'</u> of smooth muscle in bloodvessel walls are essential.
(e.g. if too many vessels are dilated at any one time adequate venous return to the heart will not be maintained.)

<u>ACTION of VALVES</u>
Once the column of blood has moved up *UNIDIRECTIONAL VALVES* in larger veins prevent backflow.

CAPILLARIES

ARTERIAL PRESSURE

VENULES

WATER BALANCE

Water makes up about 70% of adult human body, i.e. about 46 litres in 70 kg. man. Some is in Blood and Tissue Fluids; a great deal in cells themselves. In health the total amount of body water (and salt) is kept reasonably constant in spite of wide fluctuations in daily intake.

A BALANCE is struck between

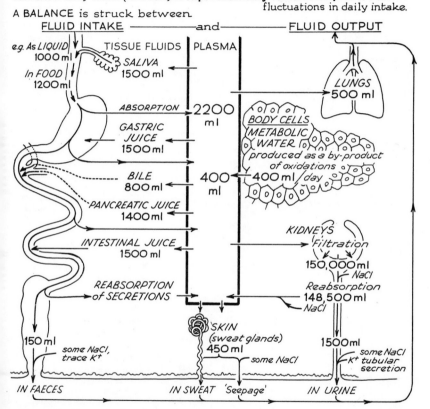

FLUID INTAKE ——— and ——— FLUID OUTPUT

e.g. As LIQUID 1000 ml
In FOOD 1200 ml

TISSUE FLUIDS | PLASMA

SALIVA 1500 ml
ABSORPTION 2200 ml
GASTRIC JUICE 1500 ml
BILE 800 ml — 400 ml
PANCREATIC JUICE 1400 ml
INTESTINAL JUICE 1500 ml
REABSORPTION of SECRETIONS

LUNGS 500 ml

BODY CELLS
METABOLIC WATER
produced as a by-product of oxidations
400 ml/day

KIDNEYS
Filtration
150,000 ml NaCl
Reabsorption
148,500 ml NaCl

150 ml
some NaCl, trace K⁺

SKIN (sweat glands) 450 ml
some NaCl

1500 ml
some NaCl K⁺ tubular secretion

IN FAECES | IN SWEAT 'Seepage' | IN URINE

Except in Growth, Convalescence or Pregnancy, when new tissue is being formed, an INCREASE or DECREASE in INTAKE leads to an appropriate INCREASE or DECREASE in OUTPUT to maintain the BALANCE.

BLOOD

Blood is the specialized fluid tissue of the *TRANSPORT SYSTEM*.
[Specific Gravity, 1·055–1·065; pH, 7·3–7·4; Average amount, 5 litres, varies with body weight (about 7·7% of body weight).

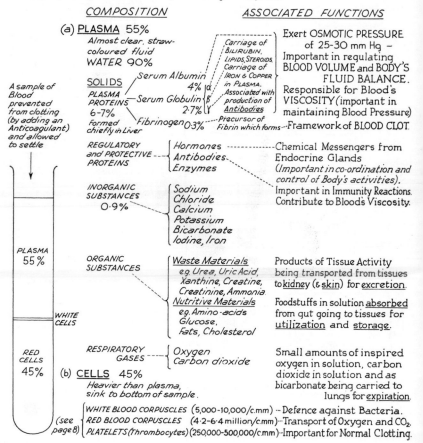

COMPOSITION

(a) PLASMA 55%
Almost clear, straw-coloured fluid
WATER 90%

SOLIDS
PLASMA PROTEINS 6–7%
formed chiefly in Liver

- Serum Albumin 4%
- Serum Globulin 2·7%
- Fibrinogen 0·3%

REGULATORY and PROTECTIVE PROTEINS
- Hormones
- Antibodies
- Enzymes

INORGANIC SUBSTANCES 0·9%
- Sodium
- Chloride
- Calcium
- Potassium
- Bicarbonate
- Iodine, Iron

ORGANIC SUBSTANCES
- Waste Materials e.g. Urea, Uric Acid, Xanthine, Creatine, Creatinine, Ammonia
- Nutritive Materials e.g. Amino-acids Glucose, Fats, Cholesterol

RESPIRATORY GASES
- Oxygen
- Carbon dioxide

(b) CELLS 45%
Heavier than plasma, sink to bottom of sample.

A sample of Blood prevented from clotting (by adding an Anticoagulant) and allowed to settle

PLASMA 55%

WHITE CELLS

RED CELLS 45%

ASSOCIATED FUNCTIONS

Carriage of BILIRUBIN, LIPIDS, STEROIDS, Carriage of IRON & COPPER in Plasma.
Associated with production of Antibodies
Precursor of Fibrin which forms Framework of BLOOD CLOT.

Exert OSMOTIC PRESSURE of 25–30 mm Hg – Important in regulating BLOOD VOLUME and BODY'S FLUID BALANCE. Responsible for Blood's VISCOSITY (important in maintaining Blood Pressure)

Chemical Messengers from Endocrine Glands (Important in co-ordination and control of Body's activities). Important in Immunity Reactions. Contribute to Blood's Viscosity.

Products of Tissue Activity being transported from tissues to kidney (& skin) for excretion.

Foodstuffs in solution absorbed from gut going to tissues for utilization and storage.

Small amounts of inspired oxygen in solution, carbon dioxide in solution and as bicarbonate being carried to lungs for expiration.

(see page 8)
- WHITE BLOOD CORPUSCLES (5,000–10,000/c.mm) – Defence against Bacteria.
- RED BLOOD CORPUSCLES (4·2–6·4 million/c.mm) – Transport of Oxygen and CO_2.
- PLATELETS (thrombocytes) (250,000–500,000/c.mm) – Important for Normal Clotting.

60

BLOOD COAGULATION

Blood does not normally clot within healthy blood vessels.
When blood is shed it undergoes a series of changes which end in <u>CLOTTING</u>.

The <u>CLASSICAL THEORY of MORAWITZ</u> recognized the interaction of <u>4 FACTORS</u>

THROMBOPLASTIN(III) —— acting on —— **PROTHROMBIN**(II)

*An enzyme thought to
be liberated by
damaged cells*

*A plasma protein— the
precursor of Thrombin-
formed in liver*

Modern Views recognize *2 separate sources of
THROMBOPLASTIN* (III) in body - each results from a
complex chain of reactions involving a number
of *FACTORS.*

in presence of
CALCIUM(IV)
*which maintains
suitable medium
in which coagulation
can take place*

In BLOOD PLASMA	*In TISSUES*
Hageman factor	*Factor V*
Plasma thromboplastin antecedent	
Factor V	*Tissue damage*
Factor VIII (Antihaemophilic globulin)	
Platelets	*Factor VII*
Factor IX (Christmas factor)	
Factor X (Stuart-Prower factor)	*Factor X (Stuart-Prower factor)*
Calcium	*Calcium*
INTRINSIC THROMBOPLASTIN	**EXTRINSIC THROMBOPLASTIN**

(I)
THROMBIN which acts on **FIBRINOGEN**
(A SOLUBLE PROTEIN FORMED IN LIVER)

Insoluble FIBRIN threads
form
<u>FRAMEWORK of BLOOD CLOT</u>

Platelets cling to inter-
sections of fibrin threads

*Both these sources of
Thromboplastin appear
to be necessary for*
<u>PHYSIOLOGICAL HAEMOSTASIS</u>
(i.e. cessation of bleeding when wound occurs) —— *also
requires* ——> <u>ADHESION and
SHRINKAGE of CLOT</u>

<u>EXPRESSION of SERUM</u>

5

FACTORS REQUIRED for NORMAL HAEMOPOIESIS

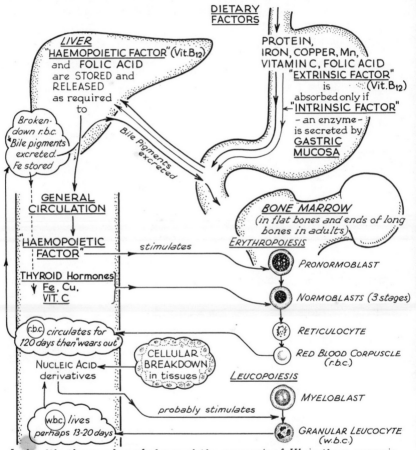

DIETARY FACTORS

LIVER
"HAEMOPOIETIC FACTOR" (Vit.B₁₂) and FOLIC ACID are STORED and RELEASED as required to

PROTEIN, IRON, COPPER, Mn, VITAMIN C, FOLIC ACID "EXTRINSIC FACTOR" is (Vit.B₁₂) absorbed only if "INTRINSIC FACTOR" — an enzyme — is secreted by: GASTRIC MUCOSA

Broken-down r.b.c. Bile pigments excreted. Fe stored

Bile Pigments excreted

GENERAL CIRCULATION

BONE MARROW (in flat bones and ends of long bones in adults)

"HAEMOPOIETIC FACTOR" — *stimulates* →

ERYTHROPOIESIS

PRONORMOBLAST

THYROID Hormones ↓ Fe, Cu, VIT. C

NORMOBLASTS (3 stages)

r.b.c. circulates for 120 days then "wears out"

RETICULOCYTE

RED BLOOD CORPUSCLE (r.b.c.)

NUCLEIC ACID derivatives ← CELLULAR BREAKDOWN in tissues

LEUCOPOIESIS

MYELOBLAST

probably stimulates →

w.b.c. lives perhaps 13-20 days

GRANULAR LEUCOCYTE (w.b.c.)

In health the number of r.b.c. and the amount of Hb in them remain fairly constant. Destruction of old red cells is balanced by formation of new.

BLOOD GROUPS

There are present in the PLASMA of some individuals substances which can cause the AGGLUTINATION *(clumping together)* and subsequent HAEMOLYSIS *(breakdown)* of the RED BLOOD CELLS of some other individuals.

If such reactions follow BLOOD TRANSFUSION the two bloods are said to be INCOMPATIBLE. Small blood vessels in lungs or brain may be blocked. Haemolysis may lead to Hb in urine and eventually to kidney failure and death.

Two FACTORS are involved in an AGGLUTINATION reaction:-
An AGGLUTINOGEN present in DONOR'S Red Blood Cell } e.g. A } or B}
A specific AGGLUTININ present in RECIPIENT'S Plasma } α } or β}

Obviously no such combination occurs naturally otherwise auto-agglutination would result.

In the ABO System

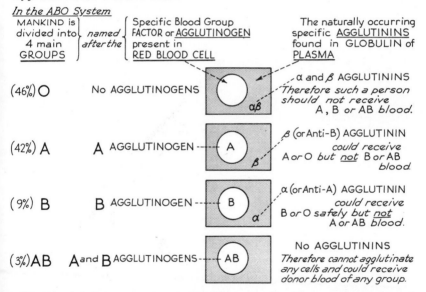

MANKIND is divided into 4 main GROUPS } *named after the* } { Specific Blood Group FACTOR or AGGLUTINOGEN present in RED BLOOD CELL	The naturally occurring specific AGGLUTININS found in GLOBULIN of PLASMA

(46%) O — No AGGLUTINOGENS — αβ
α and β AGGLUTININS
Therefore such a person should not receive A, B or AB *blood.*

(42%) A — A AGGLUTINOGEN — A — β
β (or Anti-B) AGGLUTININ
could receive A or O *but not* B or AB *blood.*

(9%) B — B AGGLUTINOGEN — B — α
α (or Anti-A) AGGLUTININ
could receive B or O *safely but not* A or AB *blood.*

(3%) AB — A and B AGGLUTINOGENS — AB
No AGGLUTININS
Therefore cannot agglutinate any cells and could receive donor blood of any group.

In Practice *it is important that the* DONOR's *cells should not be agglutinated by the* RECIPIENT's *plasma. Agglutination of Recipient's cells by Donor agglutinins is less likely to occur. To avoid sub-group incompatibility Donor's blood is always matched directly with Patient's blood.*

RHESUS FACTOR

Over 50 different BLOOD GROUP FACTORS have been demonstrated. Not all are important in blood transfusion work.

The <u>RHESUS FACTOR</u> is important

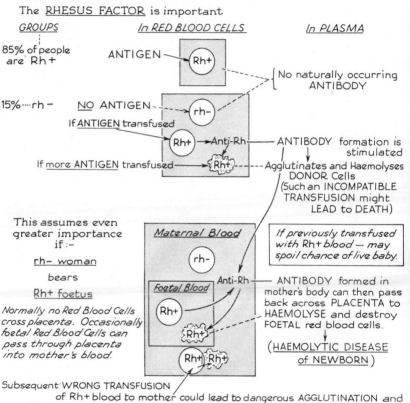

GROUPS

85% of people are Rh+

15% ···· rh −

This assumes even greater importance if :-

rh− woman

bears

Rh+ foetus

Normally no Red Blood Cells cross placenta. Occasionally foetal Red Blood Cells can pass through placenta into mother's blood.

In RED BLOOD CELLS

ANTIGEN — Rh+

<u>NO</u> ANTIGEN — rh−

If <u>ANTIGEN</u> transfused

Rh+ → Anti-Rh

If more ANTIGEN transfused → Rh+

Maternal Blood

rh−

Anti-Rh

Foetal Blood

Rh+

Rh+

Rh+ Rh+

In PLASMA

{ No naturally occurring ANTIBODY

ANTIBODY formation is stimulated

Agglutinates and Haemolyses DONOR Cells (Such an INCOMPATIBLE TRANSFUSION might LEAD to DEATH)

If previously transfused with Rh+ blood — may spoil chance of live baby.

ANTIBODY formed in mother's body can then pass back across PLACENTA to HAEMOLYSE and destroy FOETAL red blood cells.

(<u>HAEMOLYTIC DISEASE of NEWBORN</u>)

Subsequent WRONG TRANSFUSION of Rh+ blood to mother could lead to dangerous AGGLUTINATION and HAEMOLYSIS of DONOR CELLS within mother's own body.

There are many sub-groups in this system.

LYMPHATIC SYSTEM

All CELLS - - - - - - -
are bathed by TISSUE FLUID. - - - - - - -
This diffuses from CAPILLARIES. - - - - - -
Some returns to CAPILLARIES.
Some drains into blind-ending,
thin-walled LYMPHATICS. - - - - - - -
It is then known as LYMPH
(similar to plasma but less protein).

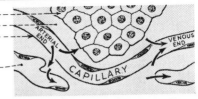

A network of LYMPHATIC VESSELS drains tissue spaces throughout the body *(except in Central Nervous System)*. They unite to form LARGER and LARGER vessels → RIGHT LYMPHATIC DUCT and LEFT THORACIC DUCT → SUBCLAVIAN VEINS *(ie. lymph is returned to the blood stream here)*

In the course of LARGER vessels, LYMPH is filtered through <u>LYMPH NODES</u>

AFFERENT LYMPHATICS – pour their LYMPH into RETICULAR FRAMEWORK of loose Sinus tissue.

MACROPHAGE cells ingest foreign material (e.g. carbon in lungs) or harmful bacteria.

LYMPH NODULES produce LYMPHOCYTES ─────→ in disintegration release PLASMA GLOBULIN (Y Globulin - associated with antibody formation and immunity reactions).

Capsule of Fibrous Tissue

EFFERENT LYMPHATIC — receives lymph after its slow passage through node.

MOVEMENT of LYMPH towards HEART depends partly on compression of lymphatic vessels by muscles of limbs and partly on 'suction' created by movements of respiration. Valves within the vessels prevent backflow. The lymphatic tissue of the body forms an important part of the Body's Defence against invading agents such as PROTOZOA, BACTERIA, VIRUSES, or their poisonous TOXINS. These act as ANTIGENS stimulating ANTIBODY FORMATION — which can subsequently destroy or neutralize the antigen.

SPLEEN

The Spleen is a vascular organ, weighing about 200 grams. It is situated in the left side of the abdomen behind the stomach and above the kidney.

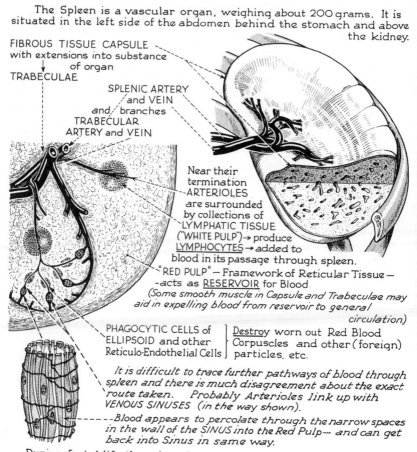

FIBROUS TISSUE CAPSULE
with extensions into substance
of organ
TRABECULAE

SPLENIC ARTERY
and VEIN
and branches
TRABECULAR
ARTERY and VEIN

Near their
termination
ARTERIOLES
are surrounded
by collections of
LYMPHATIC TISSUE
("WHITE PULP") → produce
LYMPHOCYTES → added to
blood in its passage through spleen.

"RED PULP" — Framework of Reticular Tissue —
acts as RESERVOIR for Blood.
(Some smooth muscle in Capsule and Trabeculae may
aid in expelling blood from reservoir to general
circulation)

PHAGOCYTIC CELLS of | Destroy worn out Red Blood
ELLIPSOID and other   } Corpuscles and other (foreign)
Reticulo-Endothelial Cells | particles, etc.

It is difficult to trace further pathways of blood through spleen and there is much disagreement about the exact route taken.   Probably Arterioles link up with VENOUS SINUSES (in the way shown).

Blood appears to percolate through the narrow spaces in the wall of the SINUS into the Red Pulp — and can get back into Sinus in same way.

During foetal life the spleen forms red and white blood cells.
It is not apparently essential to life in the adult.

CEREBROSPINAL FLUID

COMPOSITION:- Cerebrospinal Fluid (C.S.F.) is similar to blood plasma but does not clot. It is a clear watery alkaline fluid. It contains salts, glucose, some urea and creatinine, very little protein and very few lymphocytes.
VOLUME:- 120-150c.c. in man.   _SPECIFIC GRAVITY_:- 1·005-1·008.

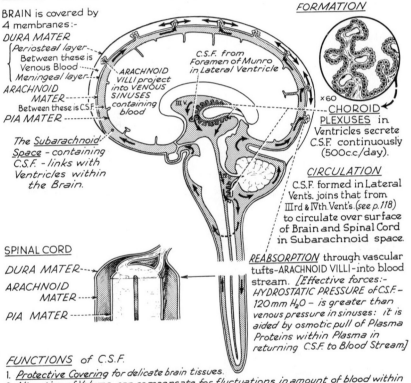

BRAIN is covered by 4 membranes:-
DURA MATER
  Periosteal layer----
  Between these is Venous Blood----
  Meningeal layer-!
ARACHNOID MATER----
  Between these is C.S.F.
PIA MATER --------

The _Subarachnoid Space_ - containing C.S.F. - links with Ventricles within the Brain.

FORMATION

C.S.F. from Foramen of Munro in Lateral Ventricle

ARACHNOID VILLI project into VENOUS SINUSES containing blood

×60

----CHOROID PLEXUSES in Ventricles secrete C.S.F. continuously (500c.c./day).

CIRCULATION
C.S.F. formed in Lateral Vent's. joins that from IIIrd & IVth.Vent's.(see p.118) to circulate over surface of Brain and Spinal Cord in Subarachnoid space.

SPINAL CORD
DURA MATER----
ARACHNOID MATER----
PIA MATER-----

REABSORPTION through vascular tufts-ARACHNOID VILLI-into blood stream. _[Effective forces:- HYDROSTATIC PRESSURE of C.S.F. - 120mm H$_2$O - is greater than venous pressure in sinuses: it is aided by osmotic pull of Plasma Proteins within Plasma in returning C.S.F. to Blood Stream]_

FUNCTIONS of C.S.F.
1. _Protective Covering_ for delicate brain tissues.
2. _Alteration of Volume_ can compensate for fluctuations in amount of blood within skull and thus keep total volume of cranial contents constant.
3. _Exchange of Metabolic substances_ between Nerve Cells and C.S.F.
              (i.e. it receives some waste products.)

CHAPTER 4.

RESPIRATORY SYSTEM

All living cells require to get OXYGEN from the fluid around them and to get rid of CARBON DIOXIDE to it.

INTERNAL RESPIRATION is the exchange of these gases between tissue cells and their fluid environment.

EXTERNAL RESPIRATION is the exchange of these gases *(oxygen and carbon dioxide)* between the body and the external environment.

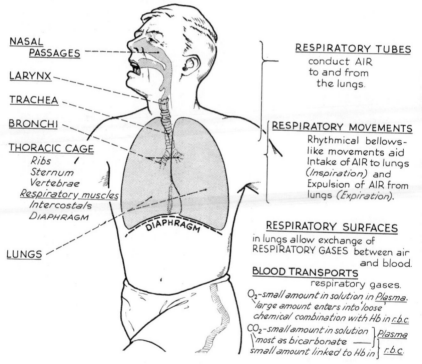

NASAL PASSAGES

LARYNX

TRACHEA

BRONCHI

THORACIC CAGE
Ribs
Sternum
Vertebrae
Respiratory muscles
Intercostals
DIAPHRAGM

LUNGS

DIAPHRAGM

RESPIRATORY TUBES
conduct AIR to and from the lungs.

RESPIRATORY MOVEMENTS
Rhythmical bellows-like movements aid Intake of AIR to lungs *(Inspiration)* and Expulsion of AIR from lungs *(Expiration)*.

RESPIRATORY SURFACES
in lungs allow exchange of RESPIRATORY GASES between air and blood.

BLOOD TRANSPORTS respiratory gases.

O_2-*small amount in solution in* <u>*Plasma*</u>.
large amount enters into loose chemical combination with Hb in <u>*r.b.c.*</u>

CO_2-*small amount in solution*] <u>*Plasma*</u>
most as bicarbonate ———]
small amount linked to Hb in] <u>*r.b.c.*</u>

AIR CONDUCTING PASSAGES

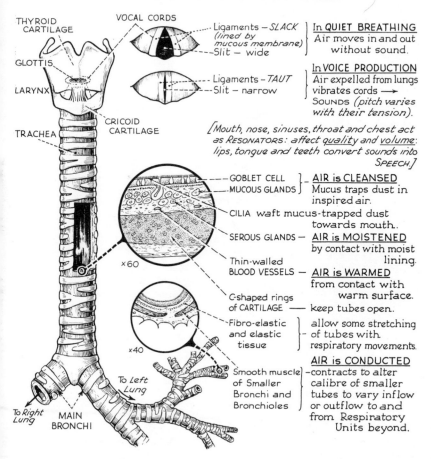

THYROID CARTILAGE

VOCAL CORDS

GLOTTIS

LARYNX

CRICOID CARTILAGE

TRACHEA

Ligaments — *SLACK (lined by mucous membrane)*
Slit — wide

Ligaments — *TAUT*
Slit — narrow

<u>In **QUIET BREATHING**</u>
Air moves in and out without sound.

<u>In **VOICE PRODUCTION**</u>
Air expelled from lungs vibrates cords ⟶ SOUNDS *(pitch varies with their tension)*.

[Mouth, nose, sinuses, throat and chest act as RESONATORS: affect quality and volume: lips, tongue and teeth convert sounds into SPEECH.]

GOBLET CELL
MUCOUS GLANDS

<u>**AIR is CLEANSED**</u>
Mucus traps dust in inspired air.

CILIA waft mucus-trapped dust towards mouth.

SEROUS GLANDS —

<u>**AIR is MOISTENED**</u>
by contact with moist lining.

Thin-walled BLOOD VESSELS —

<u>**AIR is WARMED**</u>
from contact with warm surface.

×60

C-shaped rings of CARTILAGE —— keep tubes open.

Fibro-elastic and elastic tissue

allow some stretching of tubes with respiratory movements.

×40

<u>**AIR is CONDUCTED**</u>

Smooth muscle of Smaller Bronchi and Bronchioles

contracts to alter calibre of smaller tubes to vary inflow or outflow to and from Respiratory Units beyond.

To *Left Lung*

To *Right Lung*

MAIN BRONCHI

LUNGS: RESPIRATORY SURFACES

The Trachea and the Bronchial 'Tree' conduct Air down to the
RESPIRATORY SURFACES.

There is no exchange of gases in these tubes.

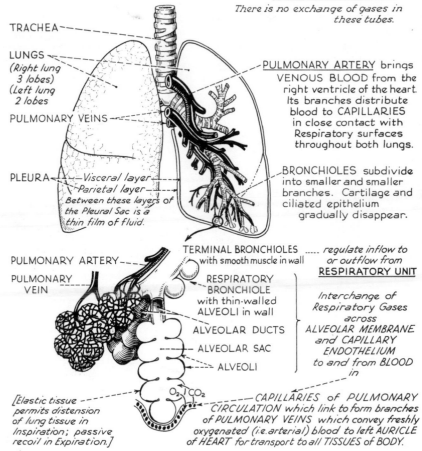

TRACHEA

LUNGS
(Right lung 3 lobes)
(Left lung 2 lobes)

PULMONARY VEINS

PLEURA — *Visceral layer*
Parietal layer
Between these layers of the Pleural Sac is a thin film of fluid.

PULMONARY ARTERY brings VENOUS BLOOD from the right ventricle of the heart. Its branches distribute blood to CAPILLARIES in close contact with Respiratory surfaces throughout both lungs.

BRONCHIOLES subdivide into smaller and smaller branches. Cartilage and ciliated epithelium gradually disappear.

PULMONARY ARTERY

PULMONARY VEIN

TERMINAL BRONCHIOLES ..... *regulate inflow to* with smooth muscle in wall *or outflow from* RESPIRATORY UNIT

RESPIRATORY BRONCHIOLE with thin-walled ALVEOLI in wall

ALVEOLAR DUCTS

ALVEOLAR SAC

ALVEOLI

Interchange of Respiratory Gases across ALVEOLAR MEMBRANE *and* CAPILLARY ENDOTHELIUM *to and from BLOOD in*

[Elastic tissue permits distension of lung tissue in Inspiration; passive recoil in Expiration.]

O_2 | CO_2

CAPILLARIES of PULMONARY CIRCULATION which link to form branches of PULMONARY VEINS which convey freshly oxygenated (i.e. arterial) blood to left AURICLE of HEART for transport to all TISSUES of BODY.

70

THORAX

The Thorax (*or* chest) is the closed cavity which contains the LUNGS, HEART and Great Vessels.

It is enclosed and bounded
ABOVE by the upper RIBS and tissues of the neck;
AT THE SIDES by the RIBS and INTERCOSTAL MUSCLES;
AT THE BACK by the RIBS and VERTEBRAL COLUMN (or *back bone*);
IN FRONT by the RIBS, COSTAL CARTILAGES and STERNUM (or *breast bone*);
BELOW by the DIAPHRAGM (*a strong dome-shaped sheet of skeletal muscle which separates the thoracic cavity from the abdominal cavity*).

It is lined by a thin moist membrane — the PLEURA — the inner layer of which invests the LUNGS. In health there is a thin film of fluid between these two pleural layers.

Capillaries of the pulmonary and not the systemic circulation supply the visceral layer. Because their blood pressure (5-10 mm Hg) is less than the osmotic 'pull' of the plasma proteins (25 mm Hg) fluid is continually 'suctioned' into the capillaries from the INTRAPLEURAL SPACE. This leads to the subatmospheric or negative intrapleural pressure essential for normal breathing.

Parietal

Visceral

Crura

DIMENSIONS of thoracic cage and the PRESSURE between pleural surfaces change rhythmically about 18-20 times a minute with the MOVEMENTS of RESPIRATION —— AIR MOVEMENT in and out of the lungs follows passively.

MECHANISM of BREATHING

The rhythmical changes in the CAPACITY of the Thorax are brought about by MUSCULAR ACTION. The changes in LUNG VOLUME with INTAKE or EXPULSION of air follow passively.

In NORMAL QUIET BREATHING

INSPIRATION

EXTERNAL INTERCOSTAL MUSCLES actively contract—
– ribs and sternum move upwards and outwards
– width of chest increases from side to side and from front to back.

DIAPHRAGM contracts—
– descends
– depth of chest increases.

CAPACITY of THORAX is INCREASED

↓

PRESSURE between PLEURAL SURFACES *(already negative)* is REDUCED from –2 to –6 mm Hg *(i.e. an increased "suction pull" is exerted on LUNG TISSUE)*

↓

ELASTIC TISSUE of LUNGS is STRETCHED

↓

LUNGS EXPAND to fill THORACIC CAVITY

↓

AIR PRESSURE within ALVEOLI is now less than atmospheric pressure

↓

AIR is sucked into ALVEOLI from ATMOSPHERE

EXPIRATION

EXTERNAL INTERCOSTAL MUSCLES relax—
– ribs and sternum move downwards and inwards
– width of chest diminishes.

DIAPHRAGM relaxes—
– ascends
– depth of chest diminishes.

CAPACITY of THORAX is DECREASED

↓

PRESSURE between PLEURAL SURFACES is INCREASED from –6 to –2 mm Hg *(i.e. less pull is exerted on LUNG TISSUE)*

↓

ELASTIC TISSUE of LUNGS RECOILS

↓

AIR PRESSURE within ALVEOLI is now greater than atmospheric pressure

↓

AIR is forced out of ALVEOLI to ATMOSPHERE

In FORCED BREATHING

Muscles of Nostrils and round Glottis may contract to aid entrance of air to lungs.
Extensors of Vertebral Column may aid inspiration.
Muscles of Neck contract—
– move 1ST rib upwards
(and sternum upwards and forwards)

Internal Intercostals may contract—
– move ribs downwards more actively.
Abdominal muscles contract—
– actively aid ascent of diaphragm.

CAPACITY of LUNGS

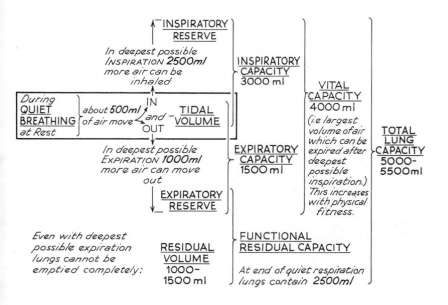

Rate and depth of breathing vary with circumstance —
e.g. amount of air moving in and out increases with muscular
activity and both inspiratory and expiratory reserves become
smaller.

At rest a normal male adult breathes in and out say 16 times per
minute. The amount of air breathed in per minute is therefore 500ml×16,
i.e. 8000 ml or 8 litres — This is the RESPIRATORY MINUTE VOLUME or
PULMONARY VENTILATION. In exercise it may go up to as much as
200 litres. These values are about 25% lower in women.

COMPOSITION of RESPIRED AIR

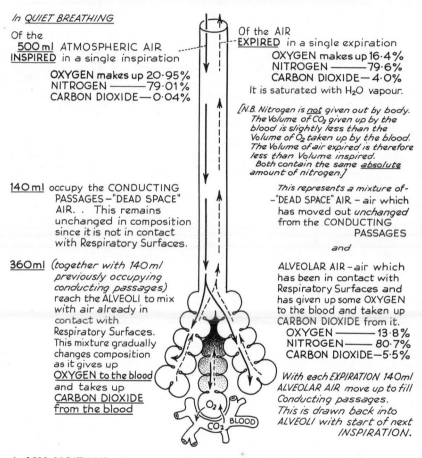

In QUIET BREATHING

Of the
500 ml ATMOSPHERIC AIR
INSPIRED *in a single inspiration*

OXYGEN makes up 20·95%
NITROGEN ———— 79·01%
CARBON DIOXIDE— 0·04%

140 ml occupy the CONDUCTING
PASSAGES—"DEAD SPACE"
AIR. . This remains
unchanged in composition
since it is not in contact
with Respiratory Surfaces.

360ml *(together with 140 ml*
previously occupying
conducting passages)
reach the ALVEOLI to mix
with air already in
contact with
Respiratory Surfaces.
This mixture gradually
changes composition
as it gives up
OXYGEN to the blood
and takes up
CARBON DIOXIDE
from the blood

Of the AIR
EXPIRED *in a single expiration*
OXYGEN makes up 16·4%
NITROGEN ———— 79·6%
CARBON DIOXIDE— 4·0%
It is saturated with H_2O vapour.

[N.B. Nitrogen is __not__ given out by body.
The Volume of CO_2 given up by the
blood is slightly less than the
Volume of O_2 taken up by the blood.
The Volume of air expired is therefore
less than Volume inspired.
Both contain the same __absolute__
amount of nitrogen.]

This represents a mixture of-
-"DEAD SPACE" AIR — air which
has moved out *unchanged*
from the CONDUCTING
PASSAGES

and

ALVEOLAR AIR –air which
has been in contact with
Respiratory Surfaces and
has given up some OXYGEN
to the blood and taken up
CARBON DIOXIDE from it.
OXYGEN ———— 13·8%
NITROGEN ——— 80·7%
CARBON DIOXIDE—5·5%

With each EXPIRATION 140ml
ALVEOLAR AIR move up to fill
Conducting passages.
This is drawn back into
ALVEOLI with start of next
INSPIRATION.

In DEEP BREATHING the composition of Air breathed out with each
expiration comes closer to that of Alveolar Air.

INTERCHANGE of RESPIRATORY GASES

A gas moves from an area where it is present at *higher concentration* to an area where it is present at *lower concentration*. The movement of gas molecules continues till the pressure exerted by them is the same throughout both areas. DRY Atmospheric Air *(at sea level)* has a pressure of 760 mm Hg.

EXPIRED AIR
    is laden with water vapour
    exerting a pressure = 47 mm Hg
    O_2, 16.4% of (760-47) = 116.2 mm Hg
    CO_2, 4% of 713 = 28.5 mm Hg

INSPIRED AIR

    O_2, 20.95% of 760 = 159 mm Hg
    CO_2, 0.04% of 760 = 0.3 mm Hg

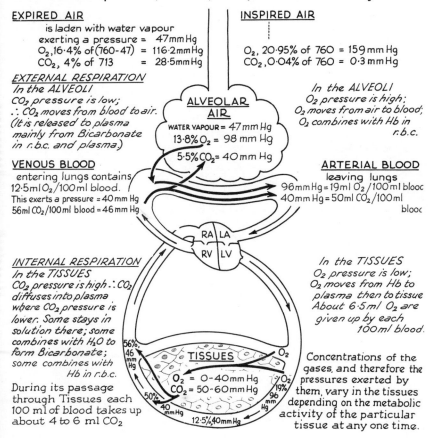

EXTERNAL RESPIRATION
In the ALVEOLI
CO_2 pressure is low;
∴ CO_2 moves from blood to air.
(It is released to plasma
mainly from Bicarbonate
in r.b.c. and plasma.)

ALVEOLAR AIR
WATER VAPOUR = 47 mm Hg
13.8% O_2 = 98 mm Hg
5.5% CO_2 = 40 mm Hg

In the ALVEOLI
O_2 pressure is high;
O_2 moves from air to blood;
O_2 combines with Hb in
r.b.c.

VENOUS BLOOD
    entering lungs contains
    12.5 ml O_2/100 ml blood.
    This exerts a pressure = 40 mm Hg
    56 ml CO_2/100 ml blood = 46 mm Hg

ARTERIAL BLOOD
    leaving lungs
    96 mm Hg = 19 ml O_2/100 ml blood
    40 mm Hg = 50 ml CO_2/100 ml blood

RA LA
RV LV

INTERNAL RESPIRATION
In the TISSUES
CO_2 pressure is high ∴ CO_2
diffuses into plasma
where CO_2 pressure is
lower. Some stays in
solution there; some
combines with H_2O to
form Bicarbonate;
some combines with
Hb in r.b.c.

During its passage
through Tissues each
100 ml of blood takes up
about 4 to 6 ml CO_2

TISSUES
O_2 = 0-40 mm Hg
CO_2 = 50-60 mm Hg

56%
46 mm Hg
50%
40 mm Hg
12.5%, 40 mm Hg
70% O_2
19%
96 mm Hg

In the TISSUES
O_2 pressure is low;
O_2 moves from Hb to
plasma then to tissue
About 6.5 ml O_2 are
given up by each
100 ml blood.

Concentrations of the
gases, and therefore the
pressures exerted by
them, vary in the tissues
depending on the metabolic
activity of the particular
tissue at any one time.

NERVOUS CONTROL of RESPIRATORY MOVEMENTS

Normal Respiratory Movements are Involuntary. They are carried out automatically (i.e. without conscious control) through the rhythmical discharge of nerve impulses from <u>CONTROLLING CENTRES in the BRAIN</u>.

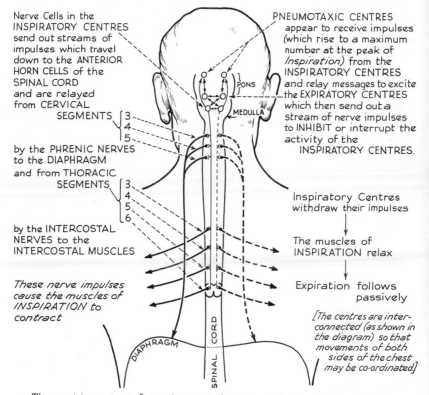

Nerve Cells in the INSPIRATORY CENTRES send out streams of impulses which travel down to the ANTERIOR HORN CELLS of the SPINAL CORD and are relayed from CERVICAL SEGMENTS [3 4 5]

by the PHRENIC NERVES to the DIAPHRAGM and from THORACIC SEGMENTS [3 4 5 6]

by the INTERCOSTAL NERVES to the INTERCOSTAL MUSCLES

These nerve impulses cause the muscles of INSPIRATION to contract

PNEUMOTAXIC CENTRES appear to receive impulses (which rise to a maximum number at the peak of *Inspiration*) from the INSPIRATORY CENTRES and relay messages to excite the EXPIRATORY CENTRES which then send out a stream of nerve impulses to INHIBIT or interrupt the activity of the INSPIRATORY CENTRES.

Inspiratory Centres withdraw their impulses

↓

The muscles of INSPIRATION relax

↓

Expiration follows passively

[The centres are interconnected (as shown in the diagram) so that movements of both sides of the chest may be co-ordinated]

PONS

MEDULLA

DIAPHRAGM

SPINAL CORD

The most important factor in regulating the activity of the respiratory centres is the level of CARBON DIOXIDE in the blood — an increase stimulates; a decrease depresses the centre.

VOLUNTARY and REFLEX FACTORS in the REGULATION of RESPIRATION

Although fundamentally automatic and regulated by the level of CARBON DIOXIDE in the blood, ingoing impulses from many parts of the body also modify the activity of the RESPIRATORY CENTRES and consequently alter the outgoing impulses to the Respiratory muscles to co-ordinate RHYTHM, RATE or DEPTH of breathing with other activities of the body.

Impulses from HIGHER CENTRES - PSYCHIC and EMOTIONAL INFLUENCES

Voluntary alterations in Breathing.
Interruptions of expiration in SPEECH and SINGING.
Deep inspiration then short spasmodic expirations in LAUGHTER and Prolonged expiration in SIGHING. WEEPING.
Deep inspiration with mouth open in YAWNING.
Slow shallow breathing in SUSPENSE and CONCENTRATION.
Rapid breathing in FEAR and EXCITEMENT.

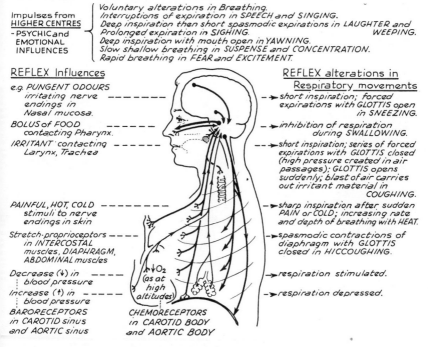

REFLEX Influences

e.g. PUNGENT ODOURS
irritating nerve endings in Nasal mucosa.

BOLUS of FOOD *contacting Pharynx.*

IRRITANT *contacting Larynx, Trachea*

PAINFUL, HOT, COLD *stimuli to nerve endings in skin*

Stretch-proprioceptors *in INTERCOSTAL muscles, DIAPHRAGM, ABDOMINAL muscles*

Decrease (↓) in *blood pressure*

Increase (↑) in *blood pressure*

BARORECEPTORS *in CAROTID sinus and AORTIC sinus*

CHEMORECEPTORS *in CAROTID BODY and AORTIC BODY*

↓O₂ *(as at high altitudes)*

REFLEX alterations in Respiratory movements

→ *short inspiration; forced expirations with GLOTTIS open in SNEEZING.*

→ *inhibition of respiration during SWALLOWING.*

→ *short inspiration; series of forced expirations with GLOTTIS closed (high pressure created in air passages); GLOTTIS opens suddenly; blast of air carries out irritant material in COUGHING.*

→ *sharp inspiration after sudden PAIN or COLD; increasing rate and depth of breathing with HEAT.*

→ *spasmodic contractions of diaphragm with GLOTTIS closed in HICCOUGHING.*

→ *respiration stimulated.*

→ *respiration depressed.*

*In Normal Breathing respiratory rate and rhythm are influenced rhythmically by the **HERING-BREUER REFLEX**.*

Distension of ALVEOLI at end of INSPIRATION	stimulates STRETCH RECEPTORS in bronchioles	Stream of ingoing impulses passes along VAGUS nerves to depress INSPIRATORY CENTRES	Withdrawal of outgoing impulses to Respiratory muscles → EXPIRATION.

CHAPTER 5.
EXCRETORY SYSTEM

Together with the *Respiratory System* and the *Skin*, the KIDNEYS are the chief *Excretory organs* of the body.

The Kidneys excrete waste products of metabolism and adjust loss of Water and Electrolytes from the body to keep body fluids relatively constant in amount and composition. They also excrete some toxic substances, e.g. many drugs.

To understand the way in which the kidney carries out these functions, it is essential to understand first the way in which it is supplied with blood. About 25% of left ventricle's output of blood in each cardiac cycle is distributed to kidneys for filtration.

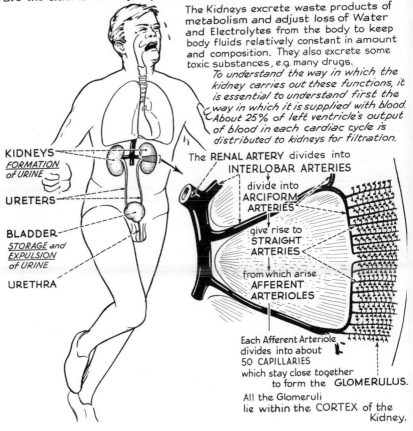

KIDNEYS
FORMATION of URINE

URETERS

BLADDER
STORAGE and *EXPULSION* of URINE

URETHRA

The RENAL ARTERY divides into INTERLOBAR ARTERIES

divide into ARCIFORM ARTERIES

give rise to STRAIGHT ARTERIES

from which arise AFFERENT ARTERIOLES

Each Afferent Arteriole divides into about 50 CAPILLARIES which stay close together to form the GLOMERULUS.

All the Glomeruli lie within the CORTEX of the Kidney.

KIDNEY

Each Kidney contains approximately one million microscopic units — *NEPHRONS* - which form URINE.

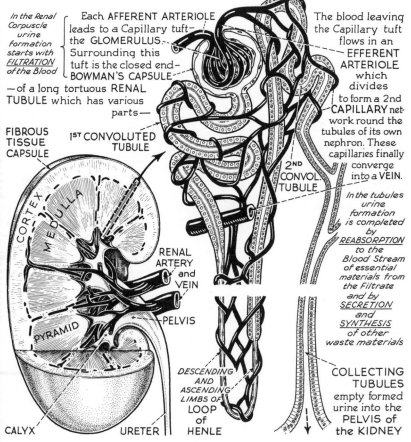

In the Renal Corpuscle urine formation starts with FILTRATION of the Blood

Each AFFERENT ARTERIOLE leads to a Capillary tuft — the GLOMERULUS. Surrounding this tuft is the closed end — BOWMAN'S CAPSULE

—of a long tortuous RENAL TUBULE which has various parts—

The blood leaving the Capillary tuft flows in an EFFERENT ARTERIOLE which divides to form a 2nd CAPILLARY network round the tubules of its own nephron. These capillaries finally converge into a VEIN.

FIBROUS TISSUE CAPSULE

1ST CONVOLUTED TUBULE

2ND CONVOL. TUBULE

In the tubules urine formation is completed by REABSORPTION to the Blood Stream of essential materials from the Filtrate and by SECRETION and SYNTHESIS of other waste materials

CORTEX

MEDULLA

PYRAMID

RENAL ARTERY and VEIN

PELVIS

CALYX

URETER

DESCENDING AND ASCENDING LIMBS OF LOOP of HENLE

COLLECTING TUBULES empty formed urine into the PELVIS of the KIDNEY

FORMATION of URINE — 1. FILTRATION

About 25% of the left ventricle's total output of blood in each cardiac cycle is distributed through the Renal Arteries to the Kidneys for FILTRATION.

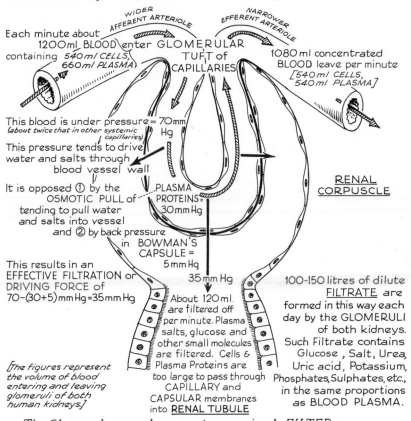

Each minute about 1200ml BLOOD enter GLOMERULAR TUFT of CAPILLARIES containing *540ml CELLS,* *660ml PLASMA*

WIDER AFFERENT ARTERIOLE

NARROWER EFFERENT ARTERIOLE

1080 ml concentrated BLOOD leave per minute [540ml CELLS, 540ml PLASMA]

This blood is under pressure = 70mm Hg *(about twice that in other systemic capillaries)*

This pressure tends to drive water and salts through blood vessel wall

It is opposed ① by the OSMOTIC PULL of PLASMA PROTEINS 30mm Hg tending to pull water and salts into vessel and ② by back pressure in BOWMAN'S CAPSULE = 5mm Hg

RENAL CORPUSCLE

This results in an EFFECTIVE FILTRATION or DRIVING FORCE of 70−(30+5)mm Hg=35mm Hg

35 mm Hg

About 120 ml. are filtered off per minute. Plasma salts, glucose and other small molecules are filtered. Cells & Plasma Proteins are too large to pass through CAPILLARY and CAPSULAR membranes into **RENAL TUBULE**

[The figures represent the volume of blood entering and leaving glomeruli of both human kidneys.]

100-150 litres of dilute FILTRATE are formed in this way each day by the GLOMERULI of both kidneys. Such Filtrate contains Glucose , Salt, Urea, Uric acid, Potassium, Phosphates, Sulphates, etc., in the same proportions as BLOOD PLASMA.

The Glomerular membrane acts as a simple FILTER — — i.e. no energy is used up by the cells in filtration.

FORMATION of URINE – 2. CONCENTRATION

As it passes along the TUBULE the FILTRATE is CONCENTRATED and essential substances are CONSERVED. The Tubular Epithelium reabsorbs water and selected materials into the blood stream. Much of this activity is regulated by hormones (see pp. 90-93,98,99). By varying the amount reabsorbed the composition of urine is adjusted to meet the body's needs at any one moment.

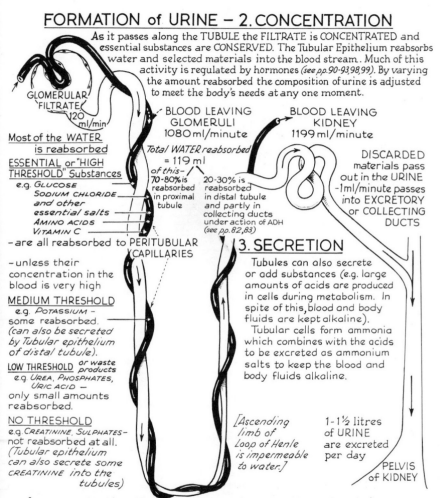

GLOMERULAR FILTRATE 120 ml/min

BLOOD LEAVING GLOMERULI 1080 ml/minute

BLOOD LEAVING KIDNEY 1199 ml/minute

Total WATER reabsorbed = 119 ml

of this –
70-80% is reabsorbed in proximal tubule

20-30% is reabsorbed in distal tubule and partly in collecting ducts under action of ADH (see pp. 82,83)

DISCARDED materials pass out in the URINE –1ml/minute passes into EXCRETORY or COLLECTING DUCTS

Most of the WATER is reabsorbed

ESSENTIAL or "HIGH THRESHOLD" Substances
e.g. GLUCOSE
SODIUM CHLORIDE and other essential salts
AMINO ACIDS
VITAMIN C
– are all reabsorbed to PERITUBULAR CAPILLARIES

– unless their concentration in the blood is very high

MEDIUM THRESHOLD
e.g. POTASSIUM – some reabsorbed. (can also be secreted by Tubular epithelium of distal tubule).

LOW THRESHOLD or waste products
e.g. UREA, PHOSPHATES, URIC ACID –
only small amounts reabsorbed.

NO THRESHOLD
e.g. CREATININE, SULPHATES – not reabsorbed at all. (Tubular epithelium can also secrete some CREATININE into the tubules)

3. SECRETION

Tubules can also secrete or add substances (e.g. large amounts of acids are produced in cells during metabolism. In spite of this, blood and body fluids are kept alkaline).

Tubular cells form ammonia which combines with the acids to be excreted as ammonium salts to keep the blood and body fluids alkaline.

[Ascending limb of Loop of Henle is impermeable to water.]

1-1½ litres of URINE are excreted per day

PELVIS of KIDNEY

In concentrating the filtrate and in secretion of certain substances, the tubular epithelial cells use energy, i.e. they do work.

REGULATION of WATER BALANCE

As the amounts of water and/or electrolytes in the body fluctuate, excretion of them is adjusted by the *Kidney* so that Body Fluids are restored to normal composition and volume.

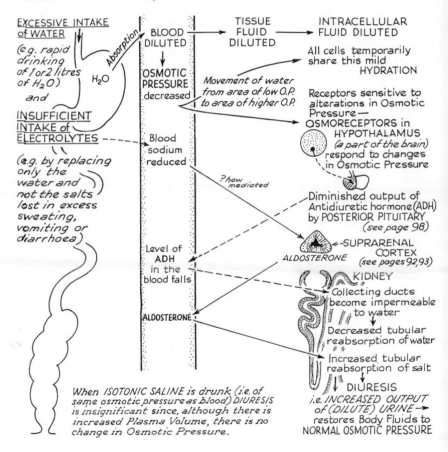

EXCESSIVE INTAKE of WATER
(e.g. rapid drinking of 1 or 2 litres of H_2O)
H_2O
and
INSUFFICIENT INTAKE of ELECTROLYTES
(e.g. by replacing only the water and not the salts lost in excess sweating, vomiting or diarrhoea)

Absorption

BLOOD DILUTED

OSMOTIC PRESSURE decreased

Blood sodium reduced

?how mediated

Level of ADH in the blood falls

ALDOSTERONE

TISSUE FLUID DILUTED

Movement of water from area of low O.P. to area of higher O.P.

INTRACELLULAR FLUID DILUTED

All cells temporarily share this mild HYDRATION

Receptors sensitive to alterations in Osmotic Pressure— OSMORECEPTORS in HYPOTHALAMUS *(a part of the brain)* respond to changes in Osmotic Pressure

Diminished output of Antidiuretic hormone (ADH) by POSTERIOR PITUITARY *(see page 98)*

ALDOSTERONE
SUPRARENAL CORTEX *(see pages 92,93)*

KIDNEY
Collecting ducts become impermeable to water

Decreased tubular reabsorption of water

Increased tubular reabsorption of salt

DIURESIS
i.e. INCREASED OUTPUT OF (DILUTE) URINE → restores Body Fluids to NORMAL OSMOTIC PRESSURE

When *ISOTONIC SALINE* is drunk (i.e. of same osmotic pressure as blood) *DIURESIS* is insignificant since, although there is increased Plasma Volume, there is no change in Osmotic Pressure.

REGULATION of WATER BALANCE

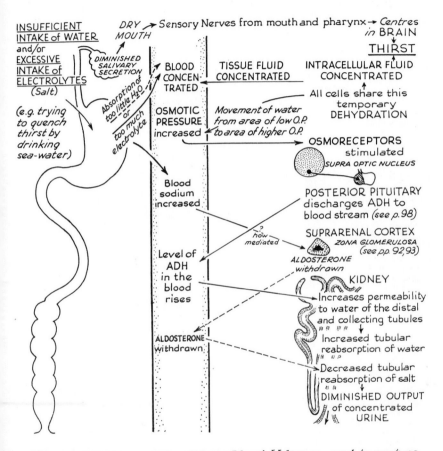

INSUFFICIENT INTAKE of WATER and/or EXCESSIVE INTAKE of ELECTROLYTES (Salt)

(e.g. trying to quench thirst by drinking sea-water)

DIMINISHED SALIVARY SECRETION

Absorption of too little H₂O or too much electrolyte

DRY MOUTH → Sensory Nerves from mouth and pharynx → Centres in BRAIN

THIRST

BLOOD CONCENTRATED — TISSUE FLUID CONCENTRATED — INTRACELLULAR FLUID CONCENTRATED

All cells share this temporary DEHYDRATION

OSMOTIC PRESSURE increased — Movement of water from area of low O.P. to area of higher O.P.

OSMORECEPTORS stimulated
SUPRA OPTIC NUCLEUS

Blood sodium increased

POSTERIOR PITUITARY discharges ADH to blood stream (see p. 98)

? how mediated

SUPRARENAL CORTEX
ZONA GLOMERULOSA (see pp. 92, 93)
ALDOSTERONE withdrawn

Level of ADH in the blood rises

KIDNEY
Increases permeability to water of the distal and collecting tubules
↓
Increased tubular reabsorption of water
↓
Decreased tubular reabsorption of salt
↓
DIMINISHED OUTPUT of concentrated URINE

ALDOSTERONE withdrawn

These measures serve to restore Blood Volume; and to restore Body Fluids to normal OSMOTIC PRESSURE.

URINE

VOLUME: *In Adult*
1000 - 1500 ml/24 hours

SPECIFIC GRAVITY   1·001 - 1·040

REACTION   Normally slightly acid —
(*pH around 6*)

} Vary with *Fluid Intake* and with *Fluid Output* from other routes — *Skin, Lungs, Gut.*
[Volume reduced during *Sleep* and *Muscular Exercise* :
Specific Gravity greater on Protein diet.]

Varies with *Diet*
[acid on ordinary mixed diet:
alkaline on vegetarian diet].

COLOUR

YELLOW due to UROCHROME pigment - probably from destruction of tissue protein.

More concentrated and DARKER in early morning — less water excreted at night but unchanged amounts of Urinary Solids.

ODOUR

AROMATIC when fresh→ AMMONIACAL on standing due to bacterial decomposition of UREA to AMMONIA.

COMPOSITION

		Grams excreted in 24 hours
WATER	96%	1000–1500

INORGANIC SUBSTANCES

Sodium	6
Chloride	7
Calcium	0·2
Potassium	2
Phosphates	1·7
Sulphates	1·8

[These figures are approximate and vary widely in healthy individuals]

ORGANIC SUBSTANCES

Urea   2%   20 - 30 — derived from breakdown of *Protein* — therefore varies with Protein in Diet.

Uric Acid   0·6 — comes from *Purine* of Food and Body Tissues.

Creatinine   1·2 — from breakdown of Body Tissues; uninfluenced by amount of dietary Protein.

Ammonia   0·5 - 0·9 — formed in Kidney from *Glutamine* brought to it by Blood Stream; varies with amounts of acid substances requiring neutralization in the Kidney.

[In the Newborn, Volume and Specific Gravity are low and Composition varies.]

URINARY BLADDER and URETERS

A resistant, distensible *Transitional Epithelium* lines all *Urinary Passages.*

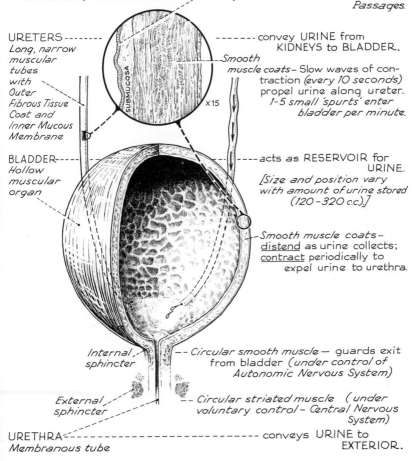

URETERS -------- ------- convey URINE from KIDNEYS to BLADDER.
Long, narrow muscular tubes with Outer Fibrous Tissue Coat and Inner Mucous Membrane

Smooth muscle coats – Slow waves of contraction *(every 10 seconds)* propel urine along ureter. *1-5 small 'spurts' enter bladder per minute.*

SUBMUCOSA   x 15

BLADDER ---- ---- acts as RESERVOIR for URINE.
Hollow muscular organ

[Size and position vary with amount of urine stored (120 - 320 c.c.).]

Smooth muscle coats – *distend* as urine collects; *contract* periodically to expel urine to urethra.

Internal sphincter -- *Circular smooth muscle* — guards exit from bladder *(under control of Autonomic Nervous System)*

External sphincter -- *Circular striated muscle (under voluntary control – Central Nervous System)*

URETHRA ---------------------------------- conveys URINE to EXTERIOR.
Membranous tube

STORAGE and EXPULSION of URINE

URINE is formed continuously by the KIDNEYS.   It collects, drop by drop, in the URINARY BLADDER which expands to hold about 300 c.c. When the Bladder is full the desire to *void urine* is experienced.

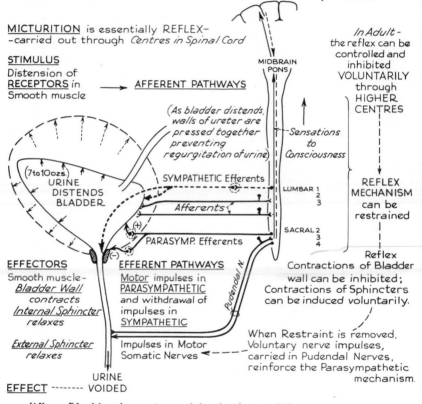

MICTURITION is essentially REFLEX—
—carried out through *Centres in Spinal Cord*

In Adult -
the reflex can be
controlled and
inhibited
VOLUNTARILY
through
HIGHER
CENTRES

STIMULUS
Distension of
RECEPTORS in
Smooth muscle

→ AFFERENT PATHWAYS

MIDBRAIN
PONS

(As bladder distends,
walls of ureter are
pressed together
preventing
regurgitation of urine)

-Sensations
to
Consciousness

(7 to 10 ozs.)
URINE
DISTENDS
BLADDER

SYMPATHETIC Efferents

LUMBAR 1
2
3

Afferents

REFLEX
MECHANISM
can be
restrained

(+)
(−)

PARASYMP. Efferents

SACRAL 2
3
4

Reflex
Contractions of Bladder
wall can be inhibited;
Contractions of Sphincters
can be induced voluntarily.

EFFECTORS
Smooth muscle-
Bladder Wall
contracts
Internal Sphincter
relaxes

External Sphincter
relaxes

EFFERENT PATHWAYS
Motor impulses in
PARASYMPATHETIC
and withdrawal of
impulses in
SYMPATHETIC

Impulses in Motor
Somatic Nerves

Pudendal N.

When Restraint is removed,
Voluntary nerve impulses,
carried in Pudendal Nerves,
reinforce the Parasympathetic
mechanism.

EFFECT ------- URINE
VOIDED

When Bladder is empty and beginning to fill −
Inhibition of Parasympathetic } *Relaxation of Bladder Wall.*
Stimulation of Sympathetic } *Constriction of Sphincters.*

CHAPTER 6.

ENDOCRINE SYSTEM

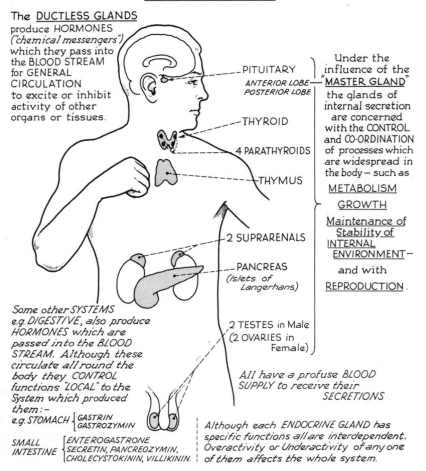

The <u>DUCTLESS GLANDS</u> produce HORMONES ("chemical messengers") which they pass into the BLOOD STREAM for GENERAL CIRCULATION to excite or inhibit activity of other organs or tissues.

PITUITARY
ANTERIOR LOBE
POSTERIOR LOBE

THYROID

4 PARATHYROIDS

THYMUS

2 SUPRARENALS

PANCREAS
(Islets of Langerhans)

2 TESTES in Male
(2 OVARIES in Female)

Under the influence of the <u>"MASTER GLAND"</u> the glands of internal secretion are concerned with the CONTROL and CO-ORDINATION of processes which are widespread in the body – such as

<u>METABOLISM</u>

<u>GROWTH</u>

<u>Maintenance of Stability of INTERNAL ENVIRONMENT</u> –

and with

<u>REPRODUCTION</u>.

All have a profuse BLOOD SUPPLY to receive their SECRETIONS

Some other SYSTEMS e.g. DIGESTIVE, also produce HORMONES which are passed into the BLOOD STREAM. Although these circulate all round the body they CONTROL functions "LOCAL" to the System which produced them :–

e.g. STOMACH { GASTRIN GASTROZYMIN

SMALL INTESTINE { ENTEROGASTRONE SECRETIN, PANCREOZYMIN, CHOLECYSTOKININ, VILLIKININ.

Although each ENDOCRINE GLAND has specific functions all are interdependent. Overactivity or Underactivity of any one of them affects the whole system.

THYROID

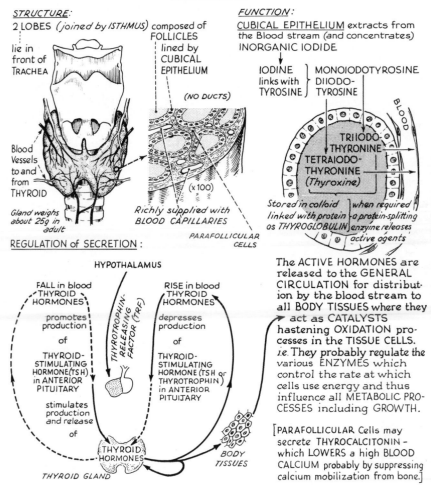

STRUCTURE:

2 LOBES *(joined by ISTHMUS)* composed of
               FOLLICLES

lie in
front of
TRACHEA
          lined by
          CUBICAL
          EPITHELIUM

(NO DUCTS)

Blood
Vessels
to and
from
THYROID

*Gland weighs
about 25g in
adult*

(x 100)

*Richly supplied with
BLOOD CAPILLARIES*

PARAFOLLICULAR
CELLS

FUNCTION:

CUBICAL EPITHELIUM extracts from
the Blood stream (and concentrates)
INORGANIC IODIDE

IODINE
links with
TYROSINE
} MONOIODOTYROSINE
  DIIODO-
  TYROSINE

BLOOD

TRIIODO-
THYRONINE
TETRAIODO-
THYRONINE
(Thyroxine)

Stored in colloid | *when required*
linked with protein | *a protein-splitting*
as THYROGLOBULIN | *enzyme releases*
               active agents

REGULATION of SECRETION :

HYPOTHALAMUS

FALL in blood
THYROID
HORMONES

promotes
production

of

THYROID-
STIMULATING
HORMONE(TSH)
in ANTERIOR
PITUITARY

stimulates
production
and release

of

THYROTROPHIN-
RELEASING
FACTOR (TRF)

RISE in blood
THYROID
HORMONES

depresses
production

of

THYROID-
STIMULATING
HORMONE (TSH or
THYROTROPHIN)
in ANTERIOR
PITUITARY

THYROID
HORMONES

BODY
TISSUES

THYROID GLAND

The ACTIVE HORMONES are
released to the GENERAL
CIRCULATION for distribution by the blood stream to
all BODY TISSUES where they
act as CATALYSTS
hastening OXIDATION pro-
cesses in the TISSUE CELLS.
i.e. They probably regulate the
various ENZYMES which
control the rate at which
cells use energy and thus
influence all METABOLIC PRO-
CESSES including GROWTH.

[PARAFOLLICULAR Cells may
secrete THYROCALCITONIN –
which LOWERS a high BLOOD
CALCIUM probably by suppressing
calcium mobilization from bone.]

THYROID

UNDERACTIVITY If the Thyroid shows *atrophy* of its secretory cells or is inadequately stimulated by the Anterior Pituitary, insufficient Hormonal Secretion is released to the blood stream and the rate at which cells use energy is reduced → Basal Metabolic Rate falls → less heat is produced → Body Temperature falls (and person feels cold). Energy stores increase (*e.g.* GLYCOGEN and FAT).

MYXOEDEMA results in the Adult

SLOWING UP OF ALL BODILY PROCESSES
Appetite is reduced; Weight increases. Gut movements are sluggish → Constipation. Heart and Respiratory Rates and Blood Pressure reduced. Thought processes slow down → Lethargy; Apathy. *SKIN* – Thick, leathery, puffy. *HAIR* – Brittle, sparse, dry. Blood cholesterol increases.

In the Child, congenital absence of the gland → *CRETINISM*

DWARFING FAILURE of SKELETAL, SEXUAL, MENTAL GROWTH and DEVELOPMENT. All "milestones" of babyhood are delayed. N.B. *Coarse skin and hair; protruding tongue*

Thyroid Extract (taken by mouth) restores individual to normal.

OVERACTIVITY If an *enlarged* Thyroid shows increased activity of its secretory cells, excess Thyroid Hormones are distributed by the blood stream to the tissues of the body → speed up oxidations in the cells, i.e. rate at which all cells use energy → the Basal Metabolic Rate is raised. As a by-product of this increased cellular activity more heat is produced → rise in Body Temperature (person feels warm).→ profuse sweating. Energy stores of body (i.e GLYCOGEN and FAT) are depleted.

SPEEDING UP OF ALL BODILY PROCESSES
{ Appetite increases but weight falls. Movements of digestive tract are increased → Diarrhoea. Heart and Respiratory Rates rise. Blood Pressure is raised. Muscular tremor and nervousness are marked. Person becomes excitable and apprehensive.

[EXOPHTHALMOS (protrusion of eyeballs) may be due to an excess of some Pituitary Hormone. It is not due to an excess of Thyroid Hormones.]

Surgical removal of part of the overactive gland reduces Thyroid activity.

PARATHYROIDS

FOUR small glands composed of *cords of cells* which secrete Parathyroid Hormone – **PARATHORMONE**
↓ or PTH
CAPILLARIES
↓
GENERAL CIRCULATION
↓
to ALL TISSUES of the body

But not all tissues are sensitive to it.

It plays an important rôle in CALCIUM and PHOSPHATE METABOLISM

PARATHYROID GLANDS

Situated behind THYROID

OESOPHAGUS

TRACHEA

Each weighs from 20-50mg in Adult

× 150

Function of EOSINOPHIL cells is unknown

PARATHORMONE acts primarily on <u>KIDNEY TUBULES</u>, <u>BONE</u> and on <u>GUT</u> to maintain BLOOD CALCIUM level at 11mg/100ml PLASMA *(necessary for normal neuromuscular excitability).*

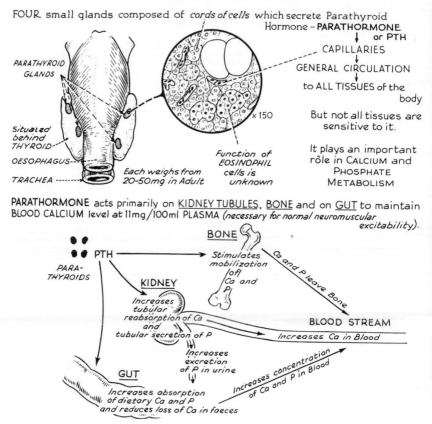

PARA-THYROIDS — PTH

BONE
Stimulates mobilization of Ca and P

Ca and P leave Bone

KIDNEY
Increases tubular reabsorption of Ca and tubular secretion of P

Increases excretion of P in urine

BLOOD STREAM
Increases Ca in Blood

GUT
Increases absorption of dietary Ca and P and reduces loss of Ca in faeces

Increases concentration of Ca and P in Blood

Alterations in concentration of CALCIUM ions in the blood are said to control Parathy-roid activity. A rise in blood Calcium depresses Parathyroid secretion. A fall in Calcium increases Parathyroid secretion. [Note: This response to the level of plasma Ca differs from thyrocalcitonin's – see 'thyroid'.] Changes in concen-tration of inorganic phosphate in blood plasma may also affect Parathyroid activity.

PARATHYROIDS

UNDERACTIVITY

– Atrophy or removal of Parathyroid tissue causes fall in BLOOD CALCIUM level and increased excitability of Neuro-muscular tissue. This leads to severe convulsive disorder — _TETANY_.
Usual manifestations:- TWITCHINGS, NERVOUSNESS, OCCASIONAL SPASMS OF FACIAL AND LIMB MUSCLES.

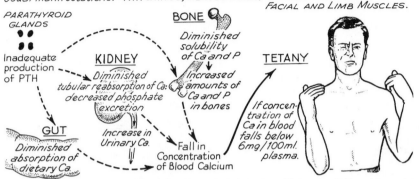

PARATHYROID GLANDS

Inadequate production of PTH

KIDNEY
Diminished tubular reabsorption of Ca; decreased phosphate excretion

Increase in Urinary Ca.

GUT
Diminished absorption of dietary Ca.

BONE
Diminished solubility of Ca and P
↓
Increased amounts of Ca and P in bones

Fall in Concentration of Blood Calcium

TETANY

If concentration of Ca in blood falls below 6mg/100ml. plasma.

Symptoms are relieved by injection of Calcium or Extract of Parathyroid

OVERACTIVITY

of the Parathyroids (_due often to tumour_) leads to rise in BLOOD CALCIUM level and eventually to _OSTEITIS FIBROSA CYSTICA_

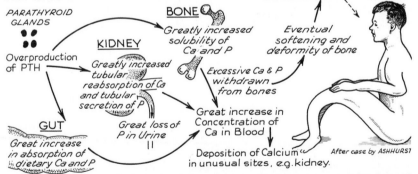

PARATHYROID GLANDS

Overproduction of PTH

KIDNEY
Greatly increased tubular reabsorption of Ca and tubular secretion of P

Great loss of P in Urine

GUT
Great increase in absorption of dietary Ca and P

BONE
Greatly increased solubility of Ca and P

Excessive Ca & P withdrawn from bones

Eventual softening and deformity of bone

Great increase in Concentration of Ca in Blood

Deposition of Calcium in unusual sites, e.g. kidney.

After case by ASHHURST

The increased level of blood calcium eventually leads to excessive loss of CALCIUM in URINE and also of WATER since the salts are excreted in solutio:
Excision of the overactive Parathyroid tissue abolishes syndrome.

SUPRARENAL GLANDS

There are TWO Suprarenal Glands. They lie close to the kidneys.

Each has an outer CORTEX and an inner MEDULLA

RIGHT KIDNEY

LEFT KIDNEY

CAPSULE
ZONA GLOMERULOSA
ZONA FASCICULATA
CORTEX
ZONA RETICULARIS
MEDULLA
×80

Secretion is under control of the ANTERIOR PITUITARY ADRENO-CORTICOTROPHIC HORMONE (CORTICOTROPHIN, ACTH)

STRESS acts via HYPOTHALAMUS

FALL in blood CORTICOIDS
promotes production of CORTICOTROPHIN
stimulates production and release of CORTICOIDS

CORTICOTROPHIN release factor

RISE in blood CORTICOIDS
depresses production of CORTICOTROPHIN

This reciprocal relationship between A.P. and S.R. Cortex leads to balanced effects on ⟶

CORTICOIDS

SUPRARENAL GLAND

The most powerful salt-retaining hormone – ALDOSTERONE – is outwith pituitary control. Its production in the Zona Glomerulosa follows a REDUCTION in BLOOD SODIUM, or a DROP in the VOLUME OF BLOOD or BODY FLUIDS (e. g. after haemorrhage, excess vomiting or diarrhoea).

The Suprarenal Cortex secretes groups of steroid HORMONES into the BLOOD STREAM. These travel to all tissues of the body.
The Suprarenal Steroids share the same functions but to varying degrees: —

1. "MINERALOCORTICOIDS" (e.g. Deoxycorticosterone and Aldosterone) have chief action on DISTAL TUBULES of the KIDNEY
They promote RETENTION of SODIUM (with water)
Stimulate EXCRETION of POTASSIUM

2. "GLUCOCORTICOIDS" (e.g. Hydrocortisone, Cortisone, Corticosterone) – on LIVER
Stimulate FORMATION of SUGAR from PROTEIN
(This gives increased BLOOD SUGAR and leads to increased GLYCOGEN stores)
1 and 2 have Anti-inflammatory and Anti-allergic properties.

3. SEX HORMONES
allied to ANDROGENS of TESTIS and OESTROGEN and PROGESTERONE of OVARY.
[Both groups are present in both sexes]
These sex hormones may influence development of SEX FUNCTIONS and of Secondary Sex characteristics.

SALT and WATER metabolism [Electrolyte balance]
CARBOHYDRATE metabolism
SEX FUNCTIONS

The Suprarenal Cortex is *essential to life* and plays an important role in states of stress.

SUPRARENAL CORTEX

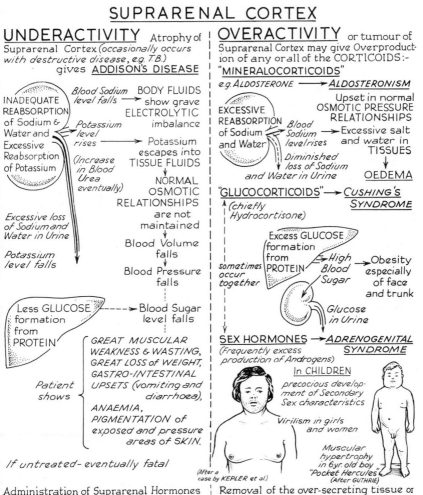

UNDERACTIVITY Atrophy of

Suprarenal Cortex (*occasionally occurs with destructive disease, e.g. T.B.*) gives **ADDISON'S DISEASE**

INADEQUATE REABSORPTION of Sodium & Water and Excessive Reabsorption of Potassium

Blood Sodium level falls → BODY FLUIDS show grave ELECTROLYTIC imbalance

Potassium level rises → Potassium escapes into TISSUE FLUIDS

(*Increase in Blood Urea eventually*)

↓ NORMAL OSMOTIC RELATIONSHIPS are not maintained
↓ Blood Volume falls
↓ Blood Pressure falls

Excessive loss of Sodium and Water in Urine

Potassium level falls

Less GLUCOSE formation from PROTEIN - - - -→ Blood Sugar level falls

Patient shows { GREAT MUSCULAR WEAKNESS & WASTING, GREAT LOSS of WEIGHT, GASTRO-INTESTINAL UPSETS (*vomiting and diarrhoea*), ANAEMIA, PIGMENTATION of exposed and pressure areas of SKIN.

If untreated - eventually fatal

Administration of Suprarenal Hormones and salt restores individual to normal.

OVERACTIVITY or tumour of

Suprarenal Cortex may give Overproduction of any or all of the CORTICOIDS:-

"MINERALOCORTICOIDS"

e.g. ALDOSTERONE ———→ *ALDOSTERONISM*

EXCESSIVE REABSORPTION of Sodium and Water

Upset in normal OSMOTIC PRESSURE RELATIONSHIPS

Blood Sodium level rises → Excessive salt and water in TISSUES

Diminished loss of Sodium and Water in Urine → OEDEMA

"GLUCOCORTICOIDS" → *CUSHING'S SYNDROME*

(chiefly Hydrocortisone)

Excess GLUCOSE formation from PROTEIN

sometimes occur together

→High Blood Sugar →Obesity especially of face and trunk

Glucose in Urine

SEX HORMONES → *ADRENOGENITAL SYNDROME*

(*Frequently excess production of Androgens*)

In CHILDREN precocious development of Secondary Sex characteristics

Virilism in girls and women

Muscular hypertrophy in 6yr. old boy "Pocket Hercules"
(After GUTHRIE)

(After a case by KEPLER et al.)

Removal of the over-secreting tissue or tumour restores individual.

SUPRARENAL MEDULLA

The Suprarenal Medulla is under control of the Hypothalamus via the Symp. Nervous System. During excitement or circumstances which demand special efforts **ADRENALINE** is released into the blood stream and prepares the various systems to react efficiently. (These effects are summed up as the *"FIGHT or FLIGHT"* FUNCTION of the Suprarenal Medullae.)

It _Constricts_ Smooth Muscle of Skin → Hairs 'stand on end'; 'Gooseflesh'.

Dilates Pupil of Eye to admit more light.

Constricts Smooth Muscle of Abdominal Blood Vessels and Cutaneous Blood Vessels → Pallor with Fright.

Dilates Smooth Muscle in Blood Vessels of Heart (Coronaries) and of Skeletal Muscles. i.e. better supply to organs requiring it in emergency.

Increases Heart Rate and Cardiac Output. Blood Pressure raised.

Relaxes Smooth Muscle in Wall of Bronchioles → better supply of air to alveoli.

Stimulates Respiration.

Inhibits Movements of Digestive Tract.

Contracts Sphincters of Gut.

Inhibits Wall of Urinary Bladder.

Contracts Ureters and Sphincter of Urinary Bladder.

Mobilizes Muscle and Liver Glycogen → increase in Blood Sugar.

Stimulates Metabolism.

Exerts favourable effect on contracting Skeletal Muscle → Fatigues less readily.

Increases Coagulability of Blood.

The Suprarenal Medullae are not essential to life – but without them the body is less able to face emergencies.

Stimulation of another part of Hypothalamus leads to release of NORADRENALINE *(from Suprarenal Medullae)* → *Vasoconstriction* → *Rise in B.P.*

ANTERIOR PITUITARY

This is the MASTER GLAND of the ENDOCRINE SYSTEM. It regulates the activity of the other Endocrine Glands, including the GONADS, and influences ALL METABOLIC PROCESSES including GROWTH.

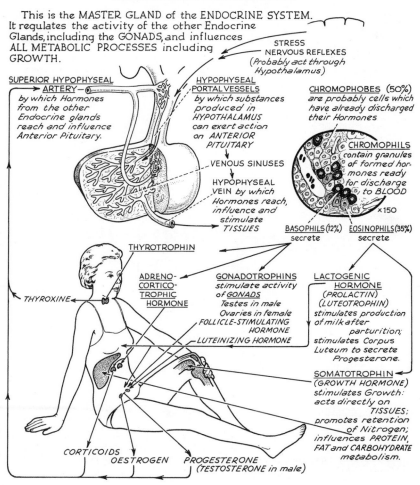

STRESS
NERVOUS REFLEXES
(Probably act through Hypothalamus)

SUPERIOR HYPOPHYSEAL
→ ARTERY—
by which Hormones from the other Endocrine glands reach and influence Anterior Pituitary.

HYPOPHYSEAL PORTAL VESSELS
by which substances produced in HYPOTHALAMUS can exert action on ANTERIOR PITUITARY

VENOUS SINUSES

HYPOPHYSEAL VEIN *by which Hormones reach, influence and stimulate TISSUES*

CHROMOPHOBES (50%) *are probably cells which have already discharged their Hormones*

CHROMOPHILS *contain granules of formed hormones ready for discharge to BLOOD*

×150

BASOPHILS (12%) secrete   EOSINOPHILS (35%) secrete

THYROTROPHIN

THYROXINE

ADRENO-CORTICO-TROPHIC HORMONE

GONADOTROPHINS *stimulate activity of GONADS Testes in male Ovaries in female* FOLLICLE-STIMULATING HORMONE LUTEINIZING HORMONE

LACTOGENIC HORMONE *(PROLACTIN) (LUTEOTROPHIN) stimulates production of milk after parturition; stimulates Corpus Luteum to secrete Progesterone.*

SOMATOTROPHIN *(GROWTH HORMONE) stimulates Growth: acts directly on TISSUES; promotes retention of Nitrogen; influences PROTEIN, FAT and CARBOHYDRATE metabolism.*

CORTICOIDS
OESTROGEN   PROGESTERONE *(TESTOSTERONE in male)*

UNDERACTIVITY of ANTERIOR PITUITARY

Deficiency or absence of **EOSINOPHIL** cells

↓

Underproduction of GROWTH Hormone (*Somatotrophin*)

LORAIN DWARF

Delayed Skeletal Growth and Retarded Sexual Development but alert, intelligent, well proportioned child.

Destructive disease of part of Anterior Pituitary(usually with damage to Posterior Pituitary and/or Hypothalamus)

↓

Underproduction of GROWTH and other ENDOCRINE-TROPHIC Hormones

FRÖHLICH'S DWARF

Stunting of Growth, Obesity (*Large appetite for sugar*); Arrested Sexual Development; Lethargic; Somnolent; Mentally Subnormal.

Atrophy of other Endocrine glands

↓

Signs of deficiency of their hormones.

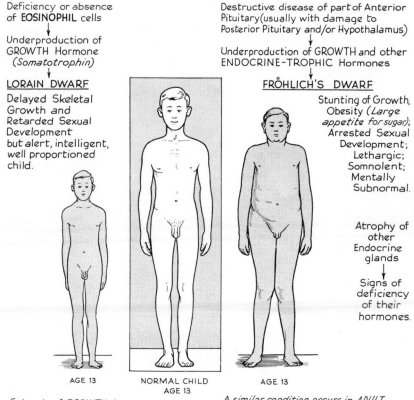

AGE 13

NORMAL CHILD AGE 13

AGE 13

Extracts of GROWTH hormone restore growth and development pattern to normal.

A similar condition occurs in ADULT without dwarfing but with suppression of sex functions and regression of secondary sex characteristics. Extracts of Growth & Gonadotrophic hormones aid in restoring patient.

Complete atrophy of all secreting cells of Anterior Pituitary in adult→<u>SIMMOND'S DISEASE</u> *(features are those of premature senility).*

OVERACTIVITY of PITUITARY CELLS

Overactivity (or tumour) of EOSINOPHILS
leads to Overproduction of GROWTH Hormone →
GIANTISM in Child; **ACROMEGALY** in Adult.

BASOPHILS
↓
Overproduction esp. of ACTH
↓ (Corticotrophin)
Overstimulation of Suprarenal Cortex
↓
CUSHING'S SYNDROME

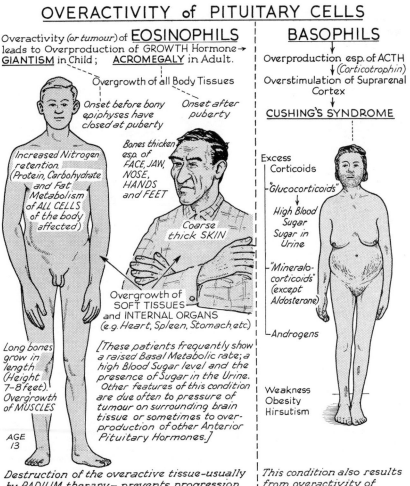

Overgrowth of all Body Tissues

Onset before bony epiphyses have closed at puberty

Onset after puberty

Increased Nitrogen retention. (Protein, Carbohydrate and Fat Metabolism of ALL CELLS of the body affected)

Bones thicken esp. of FACE, JAW, NOSE, HANDS and FEET

Coarse thick SKIN

Overgrowth of SOFT TISSUES and INTERNAL ORGANS (e.g. Heart, Spleen, Stomach, etc.)

Long bones grow in length (Height 7-8 feet). Overgrowth of MUSCLES

AGE 13

[These patients frequently show a raised Basal Metabolic rate; a high Blood Sugar level and the presence of Sugar in the Urine. Other features of this condition are due often to pressure of tumour on surrounding brain tissue or sometimes to over-production of other Anterior Pituitary Hormones.]

Excess Corticoids

—"Glucocorticoids"
↓
High Blood Sugar
Sugar in Urine

—"Mineralo-corticoids" (except Aldosterone)

—Androgens

Weakness
Obesity
Hirsutism

Destruction of the overactive tissue—usually by RADIUM therapy— prevents progression of the condition.

This condition also results from overactivity of Suprarenal Cortex itself.

POSTERIOR PITUITARY

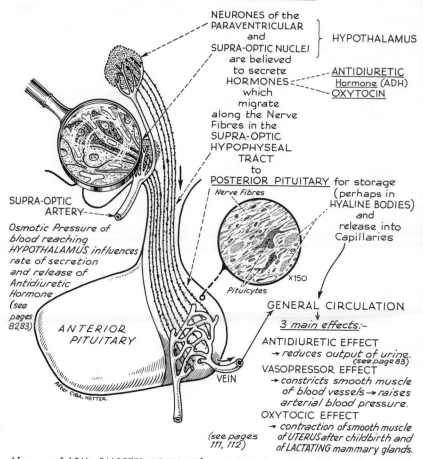

NEURONES of the PARAVENTRICULAR and SUPRA-OPTIC NUCLEI are believed to secrete HORMONES which migrate along the Nerve Fibres in the SUPRA-OPTIC HYPOPHYSEAL TRACT to POSTERIOR PITUITARY for storage (perhaps in HYALINE BODIES) and release into Capillaries

HYPOTHALAMUS

ANTIDIURETIC Hormone (ADH)
OXYTOCIN

Nerve Fibres

Pituicytes   ×150

SUPRA-OPTIC ARTERY

Osmotic Pressure of blood reaching HYPOTHALAMUS influences rate of secretion and release of Antidiuretic Hormone (see pages 82,83)

ANTERIOR PITUITARY

After CIBA: NETTER

VEIN

GENERAL CIRCULATION

3 main effects:-

ANTIDIURETIC EFFECT
→ reduces output of urine. (see page 83)

VASOPRESSOR EFFECT
→ constricts smooth muscle of blood vessels → raises arterial blood pressure.

OXYTOCIC EFFECT
→ contraction of smooth muscle of UTERUS after childbirth and of LACTATING mammary glands.

(see pages 111, 112)

Absence of ADH → DIABETES INSIPIDUS (diminished reabsorption of water from kidney filtrate → increased output of very dilute urine ; excessive thirst).

PANCREAS: ISLETS of LANGERHANS

Islets of Langerhans make up 1-2 % of Pancreatic Tissue.

x 120

α cells — secrete ⟶ GLUCAGON
(25% of Islets)

BLOOD STREAM

promotes GLYCOGEN breakdown

increases Blood Sugar and
utilization of GLUCOSE by tissues

Its effects are overshadowed by

β cells — secrete ⟶ INSULIN
(50-75% of Islets) [In cells probably in combination with Zinc]

BLOOD STREAM

SECRETION of INSULIN appears to be stimulated
primarily by a RISE in BLOOD SUGAR reaching Islets

Increases GLUCOSE UPTAKE by all
"GLUCOSE-consuming tissues" – i.e.
It plays important rôle in
CARBOHYDRATE METABOLISM,
probably regulating entrance of
GLUCOSE to cells.

ANTERIOR PITUITARY
? Indirect action on
Insulin production
GROWTH HORMONE ⎤
                 ⎬ oppose action
SUPRARENAL   ⎥ of Insulin
GLUCOCORTICOIDS ⎦

CHIEF ACTIONS on:-

ATROPHY ⟶ DIABETES
of ISLETS   MELLITUS
           complex
           disorder

Absence of
Insulin ⟶ HIGH BLOOD SUGAR ⟶
          SUGAR in URINE ⟶
          POLYURIA (since sugar lost
                    in solution in water)

THIRST ⟶ POLYDIPSIA and POLYPHAGIA

KIDNEY TUBULES — stimulates reabsorption
of GLUCOSE from FILTRATE

LIVER — reduces GLYCOGEN
breakdown ⟶ GLUCOSE ⎤
                    ⎥ BLOOD
MUSCLES increases GLYCO- ⎥ SUGAR
GEN formation   ⎥ LOWERED
from GLUCOSE ⎦

Muscles are unable to use GLUCOSE efficiently ⟶ Great weakness.

Fat metabolized instead of Carbohydrate ⟨ Fat stores depleted ⟶ Wasting and loss of weight.
                                          Ketone bodies in Blood and Urine.

If untreated ⟶ progressive ⟶ drowsiness ⟶ coma ⟶ death.

Excess Secretion of Insulin ⟶ HYPERINSULINISM ⟶ Low Blood Sugar ⟶ Reduction of
metabolism of Nervous Tissues ⟶ Giddiness ⟶ Convulsions ⟶ Death.

THYMUS

The Thymus is an irregularly-shaped organ lying behind the breast bone. It is relatively large in the child and reaches its maximum size at puberty. It closely resembles a Lymph Node.

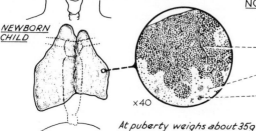

NEWBORN CHILD

×40

NODULES or FOLLICLES
contain large numbers of THYMOCYTES *(or Lymphocytes)*. These are closely packed in the outer CORTEX, less densely packed in the central MEDULLA where occasional concentric corpuscles of HASSALL are found *(function unknown)*.

At puberty weighs about 35g

The Thymus regresses after PUBERTY

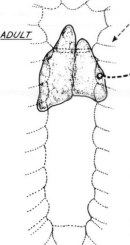

ADULT

×40

Normally the active lymphatic tissue is *replaced* by FATTY TISSUE.

Until recently the function of the thymus has been unknown. Lately shown to have an important IMMUNOLOGICAL rôle. e.g. If gland is removed at a certain early stage of development, an animal's lymphoid tissue fails to develop properly. Blood levels of lymphocytes remain low and there is deficiency in ANTIBODY formation with resultant failure to develop IMMUNE reactions to foreign protein, etc.

A thymic hormone has been postulated → blood stream → lymphoid tissue (lymph nodes; spleen; peyer's patches, etc.) → to stimulate production of normal lymphocytes and/or their antibody-forming powers.

<u>PRIMARY SEX ORGANS</u>   *produce the MALE GERM CELLS — <u>SPERMATOZOA</u>*
  TESTES (Two)   *and the MALE SEX HORMONE —*
                                                    <u>TESTOSTERONE</u>

Testosterone is responsible for
development at Puberty of :-
SECONDARY SEX ORGANS
  EPIDIDYMIS (Two) ———⎫
  VAS DEFERENS (Two) ——⎬ *transfer Spermatozoa from the Testes.*
  SEMINAL VESICLES (Two)-⎫
  PROSTATE GLAND———⎬ *secrete fluid medium for transport of Spermatozoa.*
  PENIS ————————— *transfers Spermatozoa from male to female.*

  and
appearance of
<u>SECONDARY SEX CHARACTERISTICS</u>
Laryngeal changes → Deep voice.
Pubic, Axillary and Facial hair.
Characteristic Male shape of
*   body.*

VAS DEFERENS - - -
SEMINAL VESICLE - - -
EPIDIDYMIS - - -
TESTIS - - - - -
SCROTUM - - - - -
                PENIS

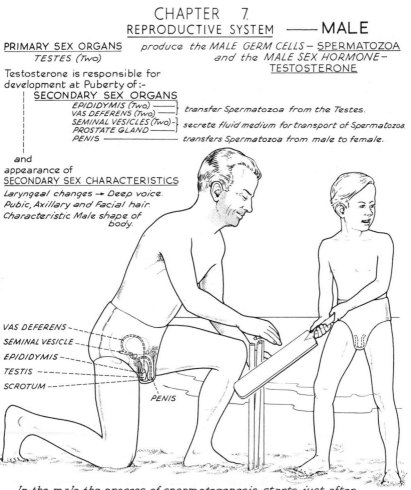

In the male the process of spermatogenesis starts just after
*   puberty and is normally continuous until old age.*

TESTIS

There are TWO TESTES.
These produce the <u>Male GERM CELLS</u>:-

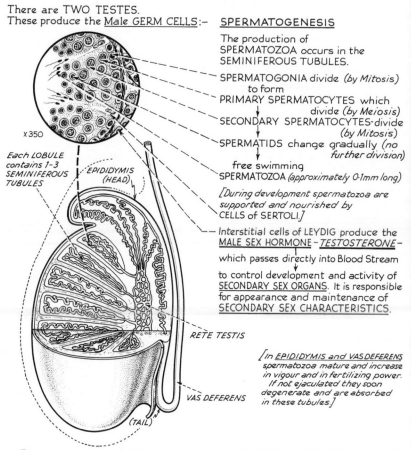

x 350

*Each LOBULE
contains 1-3
SEMINIFEROUS
TUBULES*

EPIDIDYMIS
(HEAD)

RETE TESTIS

VAS DEFERENS

(TAIL)

<u>SPERMATOGENESIS</u>

The production of
SPERMATOZOA occurs in the
SEMINIFEROUS TUBULES.

SPERMATOGONIA divide *(by Mitosis)*
to form
PRIMARY SPERMATOCYTES which
      divide *(by Meiosis)*
SECONDARY SPERMATOCYTES-divide
       (by Mitosis)
SPERMATIDS change gradually *(no
      further division)*
free swimming
SPERMATOZOA *(approximately 0·1mm long)*

*[During development spermatozoa are
supported and nourished by*
CELLS of SERTOLI.*]*

Interstitial cells of LEYDIG produce the
<u>MALE SEX HORMONE</u> – *<u>TESTOSTERONE</u>* –

which passes directly into Blood Stream

to control development and activity of
<u>SECONDARY SEX ORGANS</u>. It is responsible
for appearance and maintenance of
<u>SECONDARY SEX CHARACTERISTICS</u>.

*[In <u>EPIDIDYMIS and VAS DEFERENS</u>
spermatozoa mature and increase
in vigour and in fertilizing power.
If not ejaculated they soon
degenerate and are absorbed
in these tubules.]*

*Events occurring in the testes are under CONTROL of HORMONES chiefly
those of ANTERIOR PITUITARY.*

MALE SECONDARY SEX ORGANS

These are the organs adapted for TRANSFER of live SPERMATOZOA from male to female.

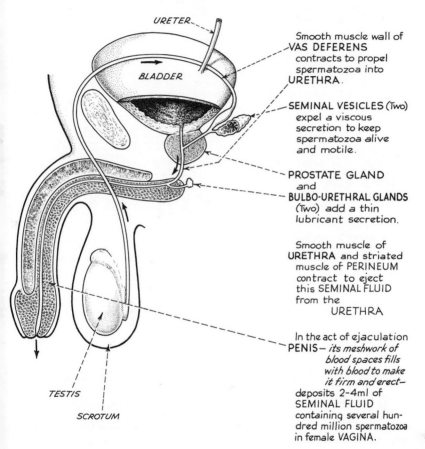

Smooth muscle wall of **VAS DEFERENS** contracts to propel spermatozoa into **URETHRA**.

SEMINAL VESICLES (Two) expel a viscous secretion to keep spermatozoa alive and motile.

PROSTATE GLAND and **BULBO-URETHRAL GLANDS** (Two) add a thin lubricant secretion.

Smooth muscle of **URETHRA** and striated muscle of **PERINEUM** contract to eject this SEMINAL FLUID from the URETHRA

In the act of ejaculation **PENIS** — *its meshwork of blood spaces fills with blood to make it firm and erect* — deposits 2–4ml of SEMINAL FLUID containing several hundred million spermatozoa in female VAGINA.

Labels on figure: URETER, BLADDER, TESTIS, SCROTUM

PUBERTY in MALE

Between the ages of 13 and 16 years Testicular tissue becomes responsive to stimulation by <u>ANTERIOR PITUITARY GONADOTROPHIC HORMONES</u>

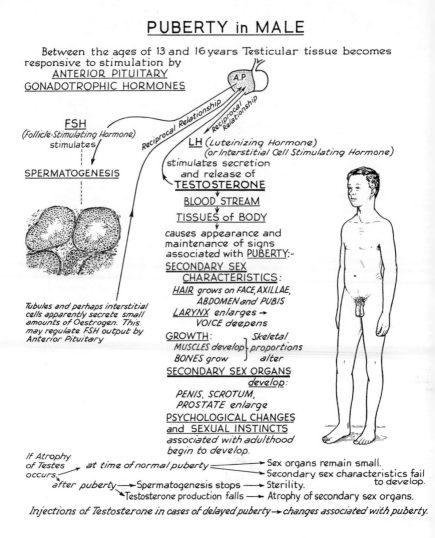

FSH
(Follicle-Stimulating Hormone)
stimulates

<u>SPERMATOGENESIS</u>

Reciprocal Relationship
Reciprocal Relationship

<u>LH</u> *(Luteinizing Hormone)*
(or Interstitial Cell Stimulating Hormone)
stimulates secretion
and release of
TESTOSTERONE

<u>BLOOD STREAM</u>

<u>TISSUES of BODY</u>

causes appearance and
maintenance of signs
associated with <u>PUBERTY</u>:-

<u>SECONDARY SEX
    CHARACTERISTICS</u>:
HAIR grows on FACE, AXILLAE,
            ABDOMEN and PUBIS
LARYNX enlarges →
            VOICE deepens

<u>GROWTH</u>:   *Skeletal*
    MUSCLES develop *proportions*
    BONES grow   *alter*

<u>SECONDARY SEX ORGANS
            develop</u>:
*PENIS, SCROTUM,
PROSTATE enlarge*

<u>PSYCHOLOGICAL CHANGES
and SEXUAL INSTINCTS</u>
*associated with adulthood
begin to develop.*

*Tubules and perhaps interstitial
cells apparently secrete small
amounts of Oestrogen. This
may regulate FSH output by
Anterior Pituitary*

*If Atrophy
of Testes
occurs* → *at time of normal puberty* → Sex organs remain small.
    → Secondary sex characteristics fail to develop.
    after puberty → Spermatogenesis stops → Sterility.
        Testosterone production falls → Atrophy of secondary sex organs.

Injections of Testosterone in cases of delayed puberty → changes associated with puberty.

REPRODUCTIVE SYSTEM — FEMALE

PRIMARY SEX ORGANS
OVARIES (Two)

produce the FEMALE GERM CELLS — OVA
and the FEMALE SEX HORMONES —
OESTROGEN *and* PROGESTERONE

Oestrogen and Progesterone are responsible
for development at Puberty of:-
SECONDARY SEX ORGANS

FALLOPIAN TUBES (Two) — *for the transfer of the Ova from Ovaries.*
VAGINA ——————— *for the reception of the Male Germ Cells.*
UTERUS ——————— *for the nutrition and development of the*
fertilized Egg Cell → developing embryo.
MAMMARY GLANDS (Two)- *for the nutrition of the New Individual after birth.*

and
appearance and
maintenance of
SECONDARY SEX
CHARACTERISTICS

Development of
Breasts

Axillary and Pubic
hair

Typical Feminine
Proportions of body

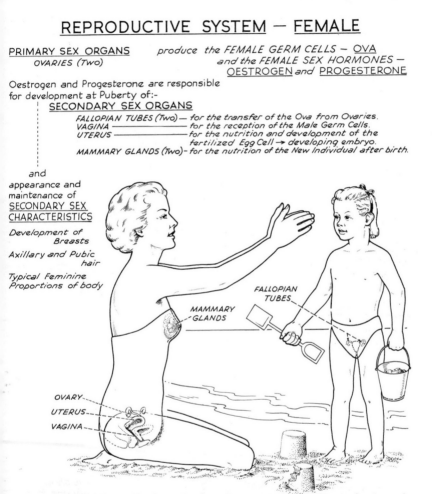

MAMMARY
GLANDS

FALLOPIAN
TUBES

OVARY
UTERUS
VAGINA

In the female the cyclical production of ova starts just after puberty and
continues (unless interrupted by pregnancy or disease) until the menopause.

105

OVARY in ORDINARY ADULT CYCLE

There are TWO OVARIES. These produce the Female GERM CELLS.
The production of OVA is a <u>CYCLICAL PROCESS</u> — <u>OOGENESIS</u>

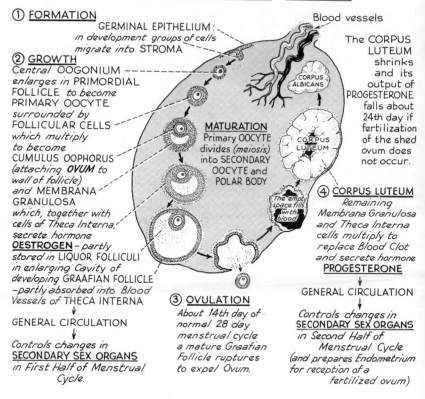

① <u>FORMATION</u>
GERMINAL EPITHELIUM:
*in development groups of cells
migrate into STROMA*

② <u>GROWTH</u>
Central OOGONIUM
enlarges in PRIMORDIAL
FOLLICLE *to become*
PRIMARY OOCYTE
surrounded by
FOLLICULAR CELLS
*which multiply
to become*
CUMULUS OOPHORUS
(attaching **OVUM** *to
wall of follicle)*
and MEMBRANA
GRANULOSA
*which, together with
cells of Theca Interna,
secrete hormone*
OESTROGEN – *partly
stored in* LIQUOR FOLLICULI
*in enlarging Cavity of
developing* GRAAFIAN FOLLICLE
– *partly absorbed into Blood
Vessels of* THECA INTERNA
↓
GENERAL CIRCULATION
↓
Controls changes in
<u>SECONDARY SEX ORGANS</u>
*in First Half of Menstrual
Cycle.*

Blood vessels

The CORPUS
LUTEUM
shrinks
and its
output of
PROGESTERONE
falls about
24th day if
fertilization
of the shed
ovum does
not occur.

CORPUS
ALBICANS

<u>MATURATION</u>
Primary OOCYTE
divides *(meiosis)*
into SECONDARY
OOCYTE and
POLAR BODY

CORPUS
LUTEUM

*The empty
space fills
with
blood*

④ <u>CORPUS LUTEUM</u>
*Remaining
Membrana Granulosa
and Theca Interna
cells multiply to
replace Blood Clot
and secrete hormone*
<u>PROGESTERONE</u>
↓
GENERAL CIRCULATION
↓
Controls changes in
<u>SECONDARY SEX ORGANS</u>
*in Second Half of
Menstrual Cycle
(and prepares Endometrium
for reception of a
fertilized ovum)*

③ <u>OVULATION</u>
*About 14th day of
normal 28 day
menstrual cycle
a mature Graafian
Follicle ruptures
to expel Ovum.*

*For simplicity the development of only one Graafian Follicle is shown here.
Several grow in each cycle but in the Human subject usually only one Follicle
ruptures. The others atrophy: i.e. ONE MATURE OVUM is shed each month.
Events in the OVARY are under control of ANTERIOR PITUITARY HORMONES.*

OVARY in PREGNANCY

When pregnancy occurs the ordinary ovarian cycle is suspended.

The CORPUS LUTEUM continues to grow until it may come to occupy 30% to 50% of the total volume of the OVARY.

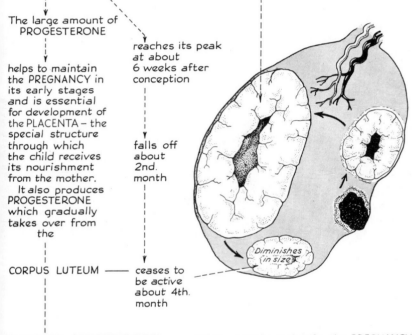

The large amount of PROGESTERONE

helps to maintain the PREGNANCY in its early stages and is essential for development of the PLACENTA – the special structure through which the child receives its nourishment from the mother.

It also produces PROGESTERONE which gradually takes over from the

CORPUS LUTEUM

reaches its peak at about 6 weeks after conception

falls off about 2nd. month

ceases to be active about 4th. month

Diminishes in size

PLACENTAL PROGESTERONE —— takes over to maintain the PREGNANCY, and to help to prepare the mammary glands for Lactation.

ADULT PELVIC SEX ORGANS
in ORDINARY FEMALE CYCLE

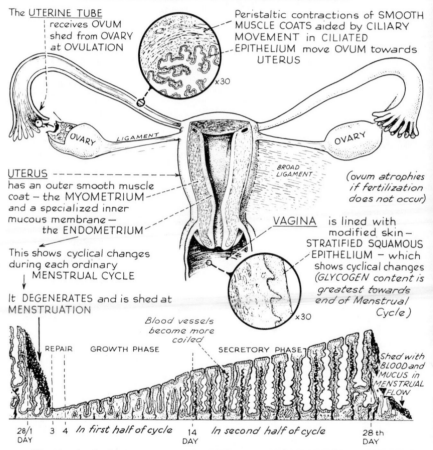

The <u>UTERINE TUBE</u> receives OVUM shed from OVARY at OVULATION

Peristaltic contractions of SMOOTH MUSCLE COATS aided by CILIARY MOVEMENT in CILIATED EPITHELIUM move OVUM towards UTERUS

×30

OVARY   LIGAMENT   OVARY

BROAD LIGAMENT

(ovum atrophies if fertilization does not occur)

<u>UTERUS</u>
has an outer smooth muscle coat — the MYOMETRIUM and a specialized inner mucous membrane — the ENDOMETRIUM

This shows cyclical changes during each ordinary MENSTRUAL CYCLE

It DEGENERATES and is shed at MENSTRUATION

<u>VAGINA</u> is lined with modified skin — STRATIFIED SQUAMOUS EPITHELIUM — which shows cyclical changes (GLYCOGEN content is greatest towards end of Menstrual Cycle)

×30

Blood vessels become more coiled

REPAIR   GROWTH PHASE   SECRETORY PHASE

Shed with BLOOD and MUCUS in MENSTRUAL FLOW

28/1 DAY   3 4   In first half of cycle   14 DAY   In second half of cycle   28th DAY

Rhythmical changes occur in UTERUS, UTERINE TUBES and VAGINA under action of OVARIAN Hormones. (see page 113)

UTERINE TUBES AND UTERUS
IN CYCLE ENDING IN PREGNANCY

The fimbriated end of the UTERINE TUBE receives OVUM at OVULATION.
The uterine tube also transmits SPERMATOZOA towards the OVA.

FERTILIZATION

I –or fusion of OVUM and SPERM– occurs in outer third of uterine tube.

Spermatozoa are of two kinds — if ovum is fertilized by one kind the resulting embryo will be MALE; if by the other, FEMALE. (All other sperms present atrophy)

The Endometrium is in LUTEAL PHASE and continues to grow. *(No menstrual degeneration occurs)* Glands are actively secreting mucus.

CLEAVAGE

After fertilization in the uterine tube the fertilized Ovum or ZYGOTE undergoes several divisions

Blastocyst

Ciliary currents and PERISTALTIC contractions in Uterine Tube carry BLASTOCYST into Uterine Secretion about 4th–7th day

Outgrowths from outer layer of Blastocyst begin to invade Endometrium

IMPLANTATION

For a few days embryo gets OXYGEN and nutrients *(by diffusion)* from the uterine glandular secretion.

Embryo sticks to lining
↓ of WOMB.
Its surface TROPHOBLAST cells fuse with, destroy and finally penetrate the ENDOMETRIUM *(now called the DECIDUA).*

Embryo now absorbs TISSUE FLUIDS and CELLULAR DEBRIS.

CHORIONIC VILLI – finger-like projections from the embryo invade mother's endometrial blood vessels.

8

109

UTERUS

PLACENTATION

PROGESTERONE is necessary for the development of the PLACENTA - the special organ through which the developing child receives nourishment from the mother.

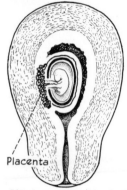

Placenta

Blood vessels develop in the Chorionic Villi which are now interlocked with mother's tissues and surrounded by mother's blood. Structure formed in this way is the PLACENTA

As PREGNANCY ADVANCES

PLACENTAL OESTROGEN, PROGESTERONE and GONADOTROPHINS in blood stream condition the maintenance of pregnancy

Umbilical Cord

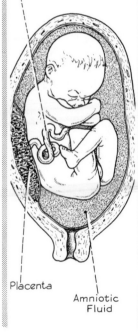

Placenta   Amniotic Fluid

PARTURITION

About 40 weeks after conception the process of CHILDBIRTH usually begins

1ST STAGE usually lasts up to 14 hours with a first birth

MYOMETRIUM
  Uterine muscle is now very greatly distended
↓
Rhythmic contractions increase in strength and frequency
↓
Press on Amniotic Fluid

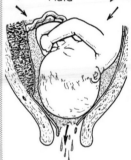

Neck of Womb (Cervix) and its opening (Os) dilate

Membranes rupture and AMNIOTIC FLUID escapes.

UTERUS

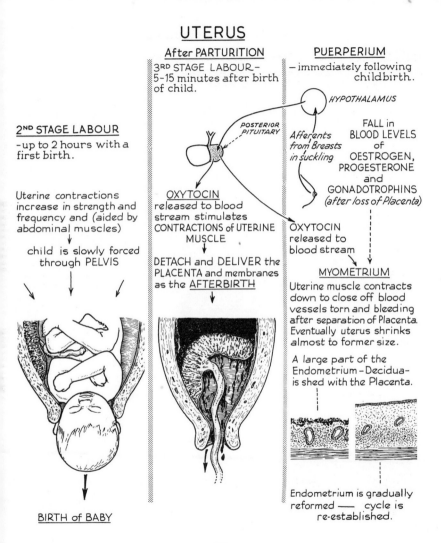

After PARTURITION

3RD STAGE LABOUR – 5-15 minutes after birth of child.

POSTERIOR PITUITARY

OXYTOCIN released to blood stream stimulates CONTRACTIONS OF UTERINE MUSCLE

DETACH and DELIVER the PLACENTA and membranes as the <u>AFTERBIRTH</u>

PUERPERIUM

– immediately following childbirth.

HYPOTHALAMUS

Afferents from Breasts in suckling

FALL in BLOOD LEVELS of OESTROGEN, PROGESTERONE and GONADOTROPHINS *(after loss of Placenta)*

OXYTOCIN released to blood stream

MYOMETRIUM

Uterine muscle contracts down to close off blood vessels torn and bleeding after separation of Placenta. Eventually uterus shrinks almost to former size.

A large part of the Endometrium – Decidua – is shed with the Placenta.

Endometrium is gradually reformed —— cycle is re-established.

2ND STAGE LABOUR

- up to 2 hours with a first birth.

Uterine contractions increase in strength and frequency and (aided by abdominal muscles)

child is slowly forced through PELVIS

<u>BIRTH of BABY</u>

111

MAMMARY GLANDS

In <u>CHILDHOOD</u> — *rudimentary ducts in fibrous tissue.*

After <u>PUBERTY</u> — *under influence of OESTROGEN,
ducts grow; PROGESTERONE → acini sprout.*

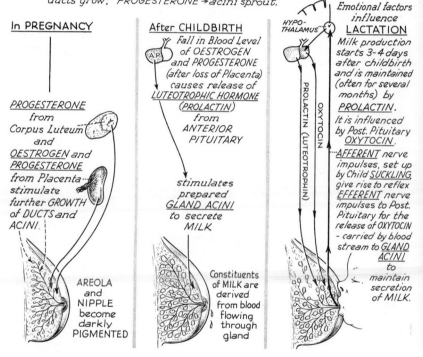

In <u>PREGNANCY</u>

<u>PROGESTERONE</u>
*from
Corpus Luteum
and*
<u>OESTROGEN</u> *and*
<u>PROGESTERONE</u>
*from Placenta
stimulate
further GROWTH
of DUCTS and
ACINI.*

AREOLA
and
NIPPLE
become
darkly
PIGMENTED

After <u>CHILDBIRTH</u>

*Fall in Blood Level
of OESTROGEN
and PROGESTERONE
(after loss of Placenta)
causes release of
<u>LUTEOTROPHIC HORMONE</u>
(PROLACTIN)
from
ANTERIOR
PITUITARY*

*stimulates
prepared
<u>GLAND ACINI</u>
to secrete
MILK*

Constituents
of MILK are
derived
from blood
flowing
through
gland

HYPO-
THALAMUS

PROLACTIN (LUTEOTROPHIN)

OXYTOCIN

*Emotional factors
influence*
LACTATION
*Milk production
starts 3-4 days
after childbirth
and is maintained
(often for several
months) by*
<u>PROLACTIN</u>.
*It is influenced
by Post. Pituitary*
<u>OXYTOCIN</u>.
<u>AFFERENT</u> *nerve
impulses, set up
by Child* <u>SUCKLING</u>,
give rise to reflex
<u>EFFERENT</u> *nerve
impulses to Post.
Pituitary for the
release of OXYTOCIN
- carried by blood
stream to <u>GLAND
ACINI</u>
to
maintain
secretion
of MILK.*

<u>POST-LACTATION</u> — *Re-establishment of OVARIAN CYCLE:
regression of secreting ACINI and
DUCT TISSUE till only little more
present than before pregnancy.*

RELATIONSHIP between ANTERIOR PITUITARY OVARIAN and ENDOMETRIAL CYCLES

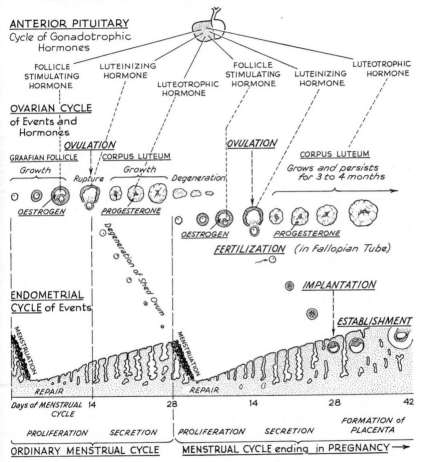

ANTERIOR PITUITARY
Cycle of Gonadotrophic Hormones

FOLLICLE STIMULATING HORMONE

LUTEINIZING HORMONE

LUTEOTROPHIC HORMONE

FOLLICLE STIMULATING HORMONE

LUTEINIZING HORMONE

LUTEOTROPHIC HORMONE

OVARIAN CYCLE
of Events and Hormones

OVULATION

GRAAFIAN FOLLICLE

Growth

Rupture

CORPUS LUTEUM

Growth

Degeneration

OVULATION

CORPUS LUTEUM

Grows and persists for 3 to 4 months

OESTROGEN

PROGESTERONE

Degeneration of Shed Ovum

OESTROGEN

PROGESTERONE

FERTILIZATION (in Fallopian Tube)

IMPLANTATION

ENDOMETRIAL CYCLE of Events

ESTABLISHMENT

MENSTRUATION

MENSTRUATION

REPAIR

REPAIR

Days of MENSTRUAL CYCLE 14   28   14   28   42

PROLIFERATION   *SECRETION*   *PROLIFERATION*   *SECRETION*

FORMATION of PLACENTA

ORDINARY MENSTRUAL CYCLE   MENSTRUAL CYCLE ending in PREGNANCY →

PUBERTY in FEMALE

Between the ages of 10 and 14 years Ovarian tissue usually begins to respond to stimulation by <u>ANTERIOR PITUITARY GONADOTROPHIC HORMONES</u>.

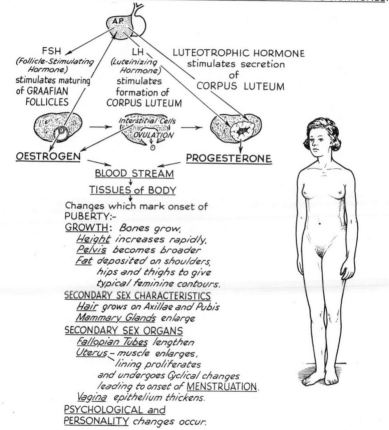

FSH
(Follicle-Stimulating Hormone)
stimulates maturing of GRAAFIAN FOLLICLES

LH
(Luteinizing Hormone)
stimulates formation of CORPUS LUTEUM

LUTEOTROPHIC HORMONE
stimulates secretion of CORPUS LUTEUM

Interstitial Cells
OVULATION

<u>OESTROGEN</u>

<u>PROGESTERONE</u>

<u>BLOOD STREAM</u>
<u>TISSUES of BODY</u>

Changes which mark onset of PUBERTY:-

<u>GROWTH</u>: *Bones grow,*
<u>*Height*</u> *increases rapidly,*
<u>*Pelvis*</u> *becomes broader*
<u>*Fat*</u> *deposited on shoulders,*
hips and thighs to give
typical feminine contours.

<u>SECONDARY SEX CHARACTERISTICS</u>
<u>*Hair*</u> *grows on Axillae and Pubis*
<u>*Mammary Glands*</u> *enlarge*

<u>SECONDARY SEX ORGANS</u>
<u>*Fallopian Tubes*</u> *lengthen*
<u>*Uterus*</u> *– muscle enlarges,*
lining proliferates
and undergoes Cyclical changes
leading to onset of <u>MENSTRUATION</u>.
<u>*Vagina*</u> *epithelium thickens.*

<u>PSYCHOLOGICAL</u> and
<u>PERSONALITY</u> *changes occur.*

i.e. PUBERTY changes GIRL CHILD into WOMAN able to bear children.

MENOPAUSE

Between the ages of 42 and 50 years OVARIAN tissue gradually ceases to respond to stimulation by <u>ANTERIOR PITUITARY GONADO-TROPHIC HORMONES.</u>

FSH
LH
LUTEOTROPHIC HORMONE

A.P.

OVARIAN CYCLE becomes irregular and finally ceases ⟶ Ovary becomes small and fibrosed and no longer produces ripe Ova.

OESTROGEN and PROGESTERONE levels in Blood stream fall.

TISSUES of the body ⟶ begin to show changes which mark the end of REPRODUCTIVE LIFE.

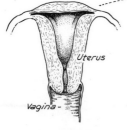

Sometimes final redistribution of fat → less typically feminine distribution.

Regression of <u>Secondary Sex Characteristics</u>.

Breasts shrink.

Ducts} Acini} Atrophy

Hair becomes sparse in axillae and on pubis.

<u>Secondary Sex Organs</u> atrophy.

Fallopian tubes shrink.

Uterine Cycle and Menstruation cease.

(Muscle and lining shrink).

Vaginal epithelium becomes thin.

External Genitalia shrink.

<u>Psychological and Personality</u> changes.

Decline in Sexual powers.

Uterus

Emotional disturbances may occur — often accompanied by Vasomotor phenomena such as "Hot Flushes" (vasodilatation), excessive sweating and giddiness.

Vagina -

After the MENOPAUSE a woman is usually unable to bear children.

CHAPTER 8.

NERVOUS AND LOCOMOTOR SYSTEMS

The Nervous System is concerned with the INTEGRATION and CONTROL of all bodily functions.

It has specialized in IRRITABILITY — *the ability to receive and respond to messages from the external and internal environments* and also in CONDUCTION — *the ability to transmit messages to and from CO-ORDINATING CENTRES.*

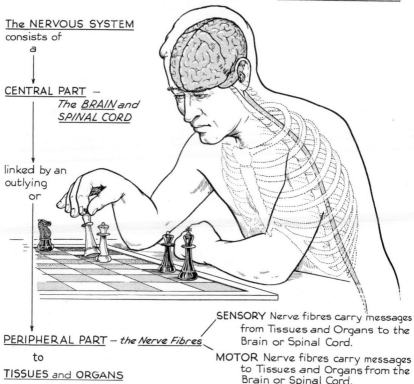

The NERVOUS SYSTEM
consists of
a
↓

CENTRAL PART —
The BRAIN and SPINAL CORD

↓

linked by an
outlying
or
↓

PERIPHERAL PART — *the Nerve Fibres*
to
TISSUES and ORGANS

SENSORY Nerve fibres carry messages from Tissues and Organs to the Brain or Spinal Cord.

MOTOR Nerve fibres carry messages to Tissues and Organs from the Brain or Spinal Cord.

CEREBRUM

The largest part of the human brain is the CEREBRUM — made up of 2 CEREBRAL HEMISPHERES. Each of these is divided into LOBES.

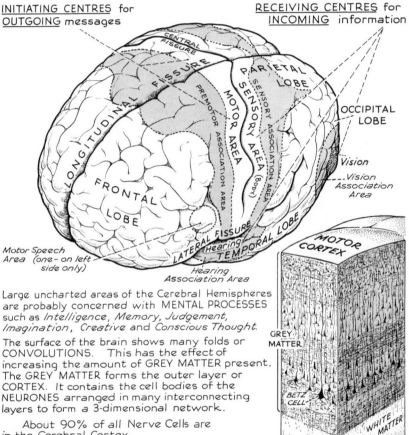

INITIATING CENTRES for OUTGOING messages

RECEIVING CENTRES for INCOMING information

CENTRAL FISSURE

LONGITUDINAL FISSURE

PREMOTOR ASSOCIATION AREA

MOTOR AREA

PARIETAL LOBE

SENSORY AREA (Body)

SENSORY ASSOCIATION AREA

OCCIPITAL LOBE

Vision

Vision Association Area

FRONTAL LOBE

LATERAL FISSURE

Hearing

TEMPORAL LOBE

Motor Speech Area (one - on left side only)

Hearing Association Area

MOTOR CORTEX

GREY MATTER

BETZ CELL

WHITE MATTER

Large uncharted areas of the Cerebral Hemispheres are probably concerned with MENTAL PROCESSES such as *Intelligence, Memory, Judgement, Imagination, Creative* and *Conscious Thought.*

The surface of the brain shows many folds or CONVOLUTIONS. This has the effect of increasing the amount of GREY MATTER present. The GREY MATTER forms the outer layer or CORTEX. It contains the cell bodies of the NEURONES arranged in many interconnecting layers to form a 3-dimensional network.

About 90% of all Nerve Cells are in the Cerebral Cortex.

HORIZONTAL SECTION through BRAIN

This view shows surface GREY MATTER containing Nerve Cells and inner WHITE MATTER made up of Nerve Fibres. Deep in the substance of the Cerebral Hemispheres there are additional masses of GREY MATTER:-

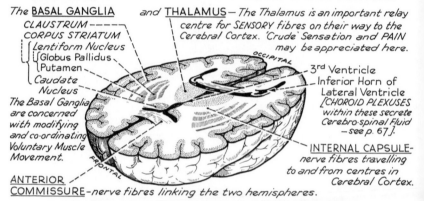

The **BASAL GANGLIA**
CLAUSTRUM - - - -
CORPUS STRIATUM
{ Lentiform Nucleus
{ { Globus Pallidus
{ { Putamen
{ Caudate
{ Nucleus

The Basal Ganglia are concerned with modifying and co-ordinating Voluntary Muscle Movement.

ANTERIOR
COMMISSURE - nerve fibres linking the two hemispheres.

and THALAMUS - The Thalamus is an important relay centre for SENSORY fibres on their way to the Cerebral Cortex. 'Crude' Sensation and PAIN may be appreciated here.

3rd Ventricle
Inferior Horn of Lateral Ventricle
[CHOROID PLEXUSES within these secrete Cerebro-spinal Fluid - see p. 67].

INTERNAL CAPSULE - nerve fibres travelling to and from centres in Cerebral Cortex.

CORONAL SECTION through BRAIN

This is a section through the TRANSVERSE (Central) FISSURE. It shows each of the major developments of the Brain:-

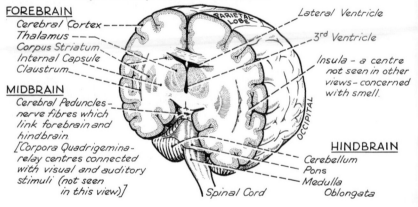

FOREBRAIN
Cerebral Cortex - - - -
Thalamus - - - -
Corpus Striatum
Internal Capsule
Claustrum

MIDBRAIN
Cerebral Peduncles - nerve fibres which link forebrain and hindbrain.
[Corpora Quadrigemina - relay centres connected with visual and auditory stimuli (not seen in this view)]

Lateral Ventricle
3rd Ventricle
Insula - a centre not seen in other views - concerned with smell.

HINDBRAIN
Cerebellum
Pons
Medulla Oblongata

Spinal Cord

VERTICAL SECTION through BRAIN

This is a Vertical Section through the LONGITUDINAL FISSURE— a deep cleft which separates the two Cerebral Hemispheres. At the bottom of this cleft are tracts of nerve fibres which link up the different LOBES of each hemisphere and also link the two hemispheres with each other — the CORPUS CALLOSUM.

FOREBRAIN

Cerebral Hemisphere

Thalamus—
-relay centres for sensation: pain appreciated here.

Hypothalamus
-contains centres for Autonomic Nervous System, e.g. Control of Heart, Blood pressure, Temperature, Metabolism, etc.

PARIETAL LOBE

OCCIPITAL LOBE

CORPUS CALLOSUM

FORNIX

FRONTAL LOBE

PONS

Pituitary gland

Opening of Lateral Ventricle

3RD Ventricle

Corpora Quadrigemina

Cerebellum
Centres concerned with balance and equilibrium. Important tracts link it with other parts of Brain and Spinal Cord.

MIDBRAIN
Receives impulses from Retina and Ear. Serves as a centre for Visual and Auditory Reflexes. In the Grey Matter are nerve cell bodies of III, IV Cranial nerves and the Red Nucleus which helps to control skilled muscular movements. The White Matter carries nerve fibres linking Red Nucleus with Cerebral Cortex, Thalamus, Cerebellum, Corpus Striatum and Spinal Cord. It also carries Ascending Sensory fibres in Lateral and Medial Lemnisci, and Descending Motor fibres on their way to Pons and Spinal Cord.

HINDBRAIN:
[PONS, CEREBELLUM, MEDULLA OBLONGATA]

Pons: Groups of Neurones form sensory nucleus of V and also nuclei of VI and VII Cranial nerves. Other nerve cells here relay impulses along their axons to Cerebellum and Cerebrum. Rubrospinal tract, Lateral and Medial Lemnisci pass through Pons and nerve fibres linking Cerebral Cortex with Medulla Oblongata and Spinal Cord.

Medulla Oblongata:
Groups of Neurones form Nuclei of VIII, IX, X, XI, XII Cranial nerves. Gracile and Cuneate Nuclei -second sensory neurones in cutaneous pathways. Tracts of Sensory fibres decussate and ascend to other side of Cerebral Cortex. Some fibres remain uncrossed. The larger part of each Motor pyramidal tract crosses and descends in other side of Spinal Cord.

CRANIAL NERVES

Twelve pairs of nerves arise directly from the undersurface of the Brain to supply Head and Neck and most of the viscera.

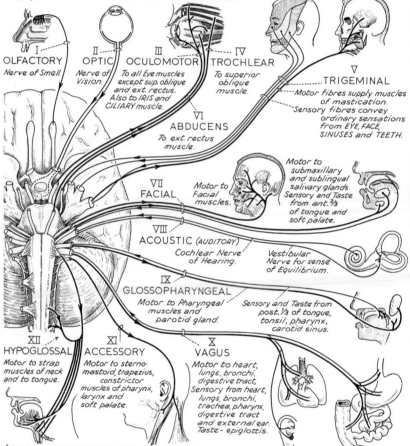

I
OLFACTORY
Nerve of Smell.

II
OPTIC
Nerve of Vision.

III
OCULOMOTOR
To all Eye muscles except sup. oblique and ext. rectus. Also to IRIS and CILIARY muscle.

IV
TROCHLEAR
To superior oblique muscle.

V
TRIGEMINAL
Motor fibres supply muscles of mastication. Sensory fibres convey ordinary sensations from EYE, FACE, SINUSES and TEETH.

VI
ABDUCENS
To ext. rectus muscle.

VII
FACIAL
Motor to Facial muscles.
Motor to submaxillary and sublingual salivary glands. Sensory and Taste from ant. ⅔ of tongue and soft palate.

VIII
ACOUSTIC *(AUDITORY)*
Cochlear Nerve of Hearing.
Vestibular Nerve for sense of Equilibrium.

IX
GLOSSOPHARYNGEAL
Motor to Pharyngeal muscles and parotid gland.
Sensory and Taste from post. ⅓ of tongue, tonsil, pharynx, carotid sinus.

XII
HYPOGLOSSAL
Motor to strap muscles of neck and to tongue.

XI
ACCESSORY
Motor to sterno-mastoid, trapezius, constrictor muscles of pharynx, larynx and soft palate.

X
VAGUS
Motor to heart, lungs, bronchi, digestive tract. Sensory from heart, lungs, bronchi, trachea, pharynx, digestive tract and external ear. Taste - epiglottis.

(After Frank H. NETTER, M.D., *The Ciba Collection of Medical Illustrations*)

SPINAL CORD

The SPINAL CORD lies within the Vertebral Canal. It is continuous above with the Medulla Oblongata.

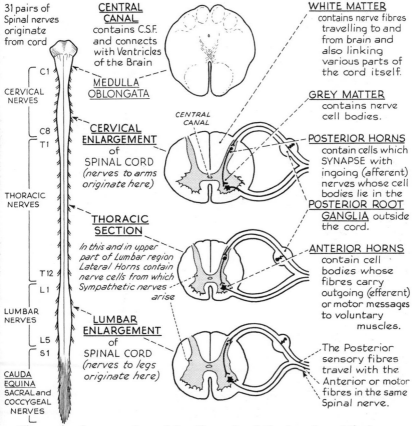

31 pairs of Spinal nerves originate from cord

CERVICAL NERVES

THORACIC NERVES

LUMBAR NERVES

CAUDA EQUINA SACRAL and COCCYGEAL NERVES

C1
C8
T1
T12
L1
L5
S1

CENTRAL CANAL
contains C.S.F. and connects with Ventricles of the Brain

MEDULLA OBLONGATA

CERVICAL ENLARGEMENT
of SPINAL CORD
(nerves to arms originate here)

CENTRAL CANAL

THORACIC SECTION
In this and in upper part of Lumbar region Lateral Horns contain nerve cells from which Sympathetic nerves arise

LUMBAR ENLARGEMENT
of SPINAL CORD
(nerves to legs originate here)

WHITE MATTER
contains nerve fibres travelling to and from brain and also linking various parts of the cord itself.

GREY MATTER
contains nerve cell bodies.

POSTERIOR HORNS
contain cells which SYNAPSE with ingoing (afferent) nerves whose cell bodies lie in the POSTERIOR ROOT GANGLIA outside the cord.

ANTERIOR HORNS
contain cell bodies whose fibres carry outgoing (efferent) or motor messages to voluntary muscles.

The Posterior sensory fibres travel with the Anterior or motor fibres in the same Spinal nerve.

The spinal nerves travel to all parts of the trunk and limbs.

121

SYNAPSE

The structural unit of the Nervous System is the NEURONE. *(see pp. 12,13)* Neurones are linked together in the Nervous System .....

..... Nerve Process to Nerve Cell body at a SYNAPSE

or

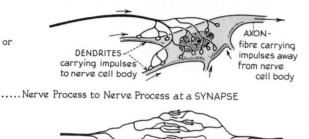

DENDRITES-
carrying impulses
to nerve cell body

AXON-
fibre carrying
impulses away
from nerve
cell body

AXON ends in small swellings – END FEET- which merely touch the DENDRITES or BODY of another NERVE CELL.
i.e. There is no direct protoplasmic union between neurones at the SYNAPSE.

..... Nerve Process to Nerve Process at a SYNAPSE

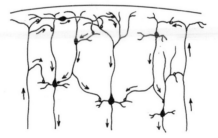

One neurone usually connects with a great many others often widely scattered in different parts of the Brain and Spinal Cord. In this way intricate chains of nerve cells forming complex pathways for incoming and outgoing information can be built up within the Central Nervous System.

When the NERVE IMPULSE – like a very small electrical current — reaches a SYNAPSE it causes the release at the NERVE ENDINGS of a CHEMICAL SUBSTANCE which bridges the gap and forms the stimulus to conduct the Impulse to the next Neurone.

A Synapse permits transmission of the impulse in one direction only.

REFLEX ACTION

The NEURONE is the ANATOMICAL or STRUCTURAL UNIT of the Nervous System: the Nervous REFLEX is the PHYSIOLOGICAL or FUNCTIONAL UNIT.

A *Nervous Reflex* is an INVOLUNTARY ACTION caused by the STIMULATION of an AFFERENT *(sensory)* nerve ending or RECEPTOR.

The structural basis of Reflex Action is the REFLEX ARC.   In its simplest form this consists of:-

RECEPTOR
in this case a muscle spindle receives a STIMULUS

AFFERENT
nerve fibre which conducts impulses

REFLEX CENTRE in the CENTRAL NERVOUS SYSTEM
(in this case the SPINAL CORD)

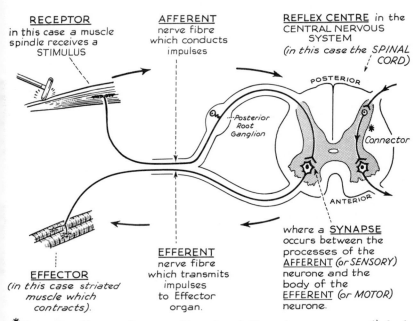

POSTERIOR

Posterior Root Ganglion

*Connector

ANTERIOR

EFFERENT
nerve fibre which transmits impulses to Effector organ.

EFFECTOR
(in this case striated muscle which contracts).

where a **SYNAPSE** occurs between the processes of the AFFERENT *(or SENSORY)* neurone and the body of the EFFERENT *(or MOTOR)* neurone.

* *In most reflex arcs in man afferent and efferent neurones are linked by at least one connector neurone.*

Reflexes form the basis of all Central Nervous System (CNS) activity. They occur at all levels of the Brain and Spinal Cord. Important bodily functions such as movements of Respiration, Digestion, etc, are all controlled through Reflexes. We are made aware of some reflex acts: others occur without our knowledge.

"EDIFICE" of the C.N.S. *(After R.C. GARRY)*

In the majority of Reflex arcs in man a chain of many connector neurones is found. There may be link-ups with various levels of the Brain and Spinal Cord.

This diagram gives a simplified concept of the LINK-UP between different levels of the Central Nervous System.

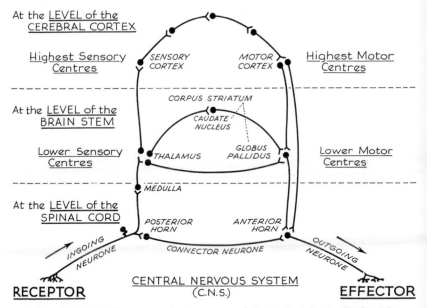

Every RECEPTOR neurone is thus potentially linked in the C.N.S with a large number of EFFECTOR organs all over the body and every EFFECTOR neurone is similarly in communication with RECEPTORS all over the body. Through such 'functional' link-ups, neurones in different parts of the Central Nervous System can influence each other. This makes it possible for 'CONDITIONED' REFLEXES to become established (for simple example see page 29). Such reflexes probably form the basis of all training so that it becomes difficult to say where REFLEX (or INVOLUNTARY) behaviour ends and purely VOLUNTARY behaviour begins.

REFLEX ACTION

Most REFLEX ACTIONS in man involve a great many REFLEX ARCS.

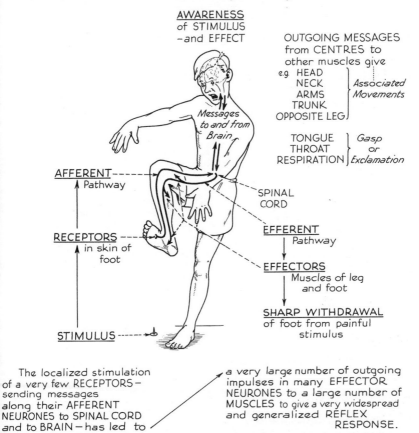

<u>AWARENESS</u>
of STIMULUS
–and EFFECT

OUTGOING MESSAGES
from CENTRES to
other muscles give
e.g. HEAD
NECK   *Associated*
ARMS   *Movements*
TRUNK
OPPOSITE LEG

TONGUE   *Gasp*
THROAT   *or*
RESPIRATION   *Exclamation*

*Messages
to and from
Brain*

<u>AFFERENT</u>
Pathway

SPINAL
CORD

<u>RECEPTORS</u>
in skin of
foot

<u>EFFERENT</u>
Pathway

<u>EFFECTORS</u>
Muscles of leg
and foot

<u>SHARP WITHDRAWAL</u>
of foot from painful
stimulus

<u>STIMULUS</u>

The localized stimulation
of a very few RECEPTORS –
sending messages
along their AFFERENT
NEURONES to SPINAL CORD
and to BRAIN – has led to

a very large number of outgoing
impulses in many EFFECTOR
NEURONES to a large number of
MUSCLES to give a very widespread
and generalized REFLEX
RESPONSE.

This is possible because *each receptor neurone is potentially connected
within the Central Nervous System with all effector neurones.*

9

SENSE ORGANS

Man's awareness of the world is limited to those forms of energy, physical or chemical, to which he has Receptors designed to respond. (Many "events" in the universe go unnoticed by man because he has no Sense Organ which can respond to them.)

Each Sense Organ is designed to respond to one type of stimulation.

EXTEROCEPTORS are stimulated by events in the EXTERNAL ENVIRONMENT.

MECHANICAL VIBRATIONS caused by object vibrating (The disturbance of molecules in atmosphere travels as waves – measured in double vibrations per second)

20,000 d.v./sec.
SOUND WAVES stimulate receptors in EAR

several 100
PRESSURE WAVES stimulate endings in SKIN
20

0 d.v./sec. Various physical or mechanical forces form CONTACT stimuli which generate nerve impulses from special sense organs in skin.

CHEMICAL CHANGES in ENVIRONMENT
Chemical substances in solution
Gaseous substances in solution

EYES
NOSE
TONGUE

stimulate HEAT receptors in skin

widespread stimulation of sense organs

ELECTROMAGNETIC WAVES (wavelengths measured in Ångström units: 1Å = one ten-millionth of a millimetre)

·00000I	COSMIC	All have chemical action on body's tissues – e.g. radiation of atomic energy 'kills' tissues.
·003	GAMMA	
I·4	X-RAYS	
250	ULTRA-VIOLET	'Sunburn'
VISIBLE RAYS 3300 7,800	INFRA-RED	2,000,000 Å (·2 mm)
	RADIO	
		1000's of METRES (KILOMETRES)
	DOMESTIC ELECTRICAL POWER WAVES	6,000,000 METRES

Exteroceptors may convey information to CONSCIOUSNESS with AWARENESS or SENSATION and lead to suitable RESPONSES planned in CEREBRAL CORTEX or they may serve as AFFERENT pathways for REFLEX (or INVOLUNTARY) ACTION with or without rising to consciousness.

PROPRIOCEPTORS are stimulated by changes in LOCOMOTOR SYSTEM of body

LABYRINTH ----- movements and position of head
MUSCLES -------- stretch
TENDONS ------- tension and stretch
JOINTS ---------- stretch and pressure

SENSE of EQUILIBRIUM or BALANCE and AWARENESS of POSITION and MOVEMENT of BODY in space.

INTEROCEPTORS in VISCERA are stimulated by changes in INTERNAL ENVIRONMENT (e.g. by distension in hollow organs).

Much of the PROPRIO- and INTEROCEPTOR information never rises to consciousness.

Overstimulation of any receptor can give rise to sensation of PAIN.

Most receptors show ADAPTATION – if continuously stimulated they send reduced numbers of impulses to the brain.

126

SMELL

Smell is a CHEMICAL SENSE, i.e. the receptors respond to CHEMICAL STIMULI. To arouse the sensation a substance must first be in a GASEOUS STATE then go into SOLUTION.

The ORGAN of SMELL is the NOSE —— *Also serves as the main air passage to Respiratory System.*

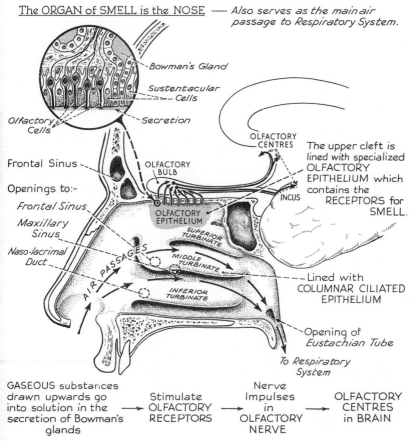

Bowman's Gland

Sustentacular Cells

Olfactory Cells

Secretion

Frontal Sinus

Openings to:-

Frontal Sinus

Maxillary Sinus

Naso-lacrimal Duct

OLFACTORY BULB

OLFACTORY EPITHELIUM

AIR PASSAGES

SUPERIOR TURBINATE

MIDDLE TURBINATE

INFERIOR TURBINATE

OLFACTORY CENTRES

The upper cleft is lined with specialized OLFACTORY EPITHELIUM which contains the RECEPTORS for SMELL.

INCUS

Lined with COLUMNAR CILIATED EPITHELIUM

Opening of *Eustachian Tube*

To Respiratory System

GASEOUS substances drawn upwards go into solution in the secretion of Bowman's glands → Stimulate OLFACTORY RECEPTORS → Nerve Impulses in OLFACTORY NERVE → OLFACTORY CENTRES in BRAIN

127

TASTE

Taste is a CHEMICAL SENSE, i.e. receptors respond to CHEMICAL STIMULI.
To arouse the sensation a substance must be in SOLUTION.

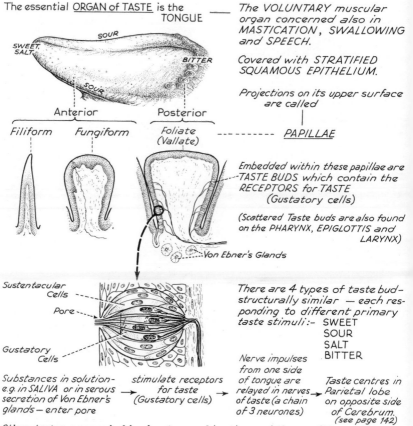

The essential <u>ORGAN of TASTE</u> is the _____ TONGUE

SOUR
SWEET SALT
BITTER
SOUR

Anterior   Posterior

Filiform   Fungiform   Foliate (Vallate)

Sustentacular Cells
Pore
Gustatory Cells

The VOLUNTARY muscular organ concerned also in MASTICATION, SWALLOWING and SPEECH.

Covered with STRATIFIED SQUAMOUS EPITHELIUM.

Projections on its upper surface are called

|

PAPILLAE

Embedded within these papillae are TASTE BUDS which contain the RECEPTORS for TASTE (Gustatory cells)

(Scattered Taste buds are also found on the PHARYNX, EPIGLOTTIS and LARYNX)

Von Ebner's Glands

There are 4 types of taste bud-structurally similar — each responding to different primary taste stimuli:-   SWEET
SOUR
SALT
BITTER

Substances in solution-
e.g. in SALIVA or in serous
secretion of Von Ebner's
glands — enter pore

→ *stimulate receptors
for taste
(Gustatory cells)*

*Nerve impulses
from one side
of tongue are
→ relayed in nerves*
*of taste (a chain
of 3 neurones)*

*Taste centres in
→ Parietal lobe
on opposite side
of Cerebrum.*
(see page 142)

*Other tastes are probably due to combinations of these with smell or
with ordinary skin sensations.*

EYE

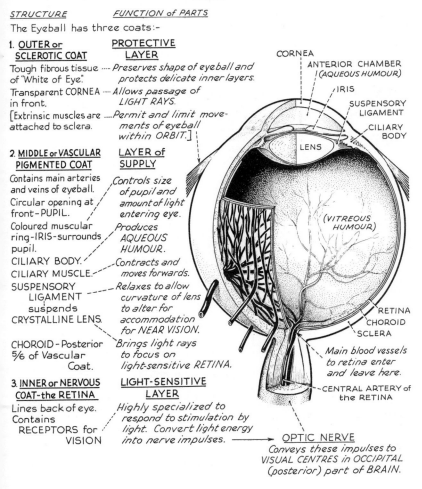

STRUCTURE

The Eyeball has three coats:-

1. OUTER or SCLEROTIC COAT

PROTECTIVE LAYER — *FUNCTION of PARTS*

Tough fibrous tissue of "White of Eye". ···· *Preserves shape of eyeball and protects delicate inner layers.*

Transparent CORNEA in front. ···· *Allows passage of LIGHT RAYS.*

[Extrinsic muscles are attached to sclera. ···· *Permit and limit movements of eyeball within ORBIT.*]

2. MIDDLE or VASCULAR PIGMENTED COAT

LAYER of SUPPLY

Contains main arteries and veins of eyeball.

Circular opening at front – PUPIL.

Coloured muscular ring-IRIS-surrounds pupil. ···· *Controls size of pupil and amount of light entering eye.*

CILIARY BODY. ···· *Produces AQUEOUS HUMOUR.*

CILIARY MUSCLE. ···· *Contracts and moves forwards.*

SUSPENSORY LIGAMENT suspends CRYSTALLINE LENS. ···· *Relaxes to allow curvature of lens to alter for accommodation for NEAR VISION.*

CHOROID – Posterior 5/6 of Vascular Coat. ···· *Brings light rays to focus on light-sensitive RETINA.*

3. INNER or NERVOUS COAT-the RETINA

LIGHT-SENSITIVE LAYER

Lines back of eye.
Contains RECEPTORS for VISION ···· *Highly specialized to respond to stimulation by light. Convert light energy into nerve impulses.* ⟶

OPTIC NERVE
Conveys these impulses to VISUAL CENTRES in OCCIPITAL (posterior) part of BRAIN.

Labels on diagram: CORNEA · ANTERIOR CHAMBER (AQUEOUS HUMOUR) · IRIS · SUSPENSORY LIGAMENT · CILIARY BODY · LENS · (VITREOUS HUMOUR) · RETINA · CHOROID · SCLERA · Main blood vessels to retina enter and leave here. · CENTRAL ARTERY of the RETINA

PROTECTION of the EYE

The hidden posterior ⅘ of the eyeball is encased in a bony socket — the ORBITAL CAVITY. A thick layer of *Areolar* and *Adipose* tissue forms a cushion between bone and eyeball. The exposed anterior ⅕ of the eyeball is protected from injury by:-

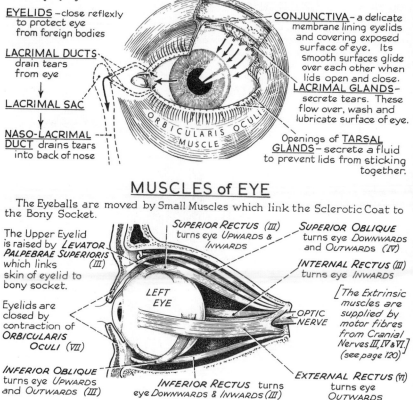

EYELIDS – close reflexly to protect eye from foreign bodies

LACRIMAL DUCTS · drain tears from eye

↓

LACRIMAL SAC

↓

NASO-LACRIMAL DUCT drains tears into back of nose

CONJUNCTIVA – a delicate membrane lining eyelids and covering exposed surface of eye. Its smooth surfaces glide over each other when lids open and close.

LACRIMAL GLANDS – secrete tears. These flow over, wash and lubricate surface of eye.

Openings of **TARSAL GLANDS** – secrete a fluid to prevent lids from sticking together.

ORBICULARIS OCULI MUSCLE

MUSCLES of EYE

The Eyeballs are moved by Small Muscles which link the Sclerotic Coat to the Bony Socket.

The Upper Eyelid is raised by *LEVATOR PALPEBRAE SUPERIORIS* (III) which links skin of eyelid to bony socket.

Eyelids are closed by contraction of *ORBICULARIS OCULI* (VII)

SUPERIOR RECTUS (III) turns eye *UPWARDS & INWARDS*

SUPERIOR OBLIQUE turns eye *DOWNWARDS and OUTWARDS (IV)*

INTERNAL RECTUS (III) turns eye *INWARDS*

LEFT EYE

OPTIC NERVE

[The Extrinsic muscles are supplied by motor fibres from Cranial Nerves III, IV & VI.] (see page 120)

INFERIOR OBLIQUE turns eye *UPWARDS and OUTWARDS (III)*

INFERIOR RECTUS turns eye *DOWNWARDS & INWARDS (III)*

EXTERNAL RECTUS (VI) turns eye *OUTWARDS*

Acting together, the Extrinsic muscles of the Eyeballs can bring about ROTATORY movements of the Eyes. Both eyes normally move together so that images fall on corresponding points of both retinae.

ACTION of LENS

The normal lens brings light rays to a sharp focus upside down on the retina. It can do this whether we are looking at an object far away or one close at hand. The curvature increases reflexly to accommodate for near vision.

Rays of light coming from every point of a DISTANT object (over 20 feet away) are PARALLEL.

The conscious mind learns to interpret the image and project it to its true position in space.

They pass through the CORNEA, AQUEOUS HUMOUR and LENS which refract them to a sharp focus — upside down and reversed from side to side — on the Retina.

Rays of light coming from a NEAR object (less than 20 feet away) RADIATE from every point

A more convex lens is required to bring these rays to a sharp focus on the Retina. Ciliary muscle contracts → Suspensory lig⁵ slackens → tension on capsule of lens slackens → lens bulges.

If the EYEBALL is <u>too short</u>, rays from a distant object are brought into focus BEHIND the Retina when the ciliary muscle is relaxed.

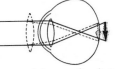

Vision is blurred.

This is longsightedness or HYPERMETROPIA.

The longsighted eye has to accommodate even for distant vision: i.e. Ciliary muscles contract to give a more convex lens and distant objects are then seen clearly. This limits amount of accommodating power left for near objects and the nearest point for sharp vision is then further away. It can be corrected by fitting spectacles with an additional convex lens.

If the EYEBALL is <u>too long</u>, rays from a distant object are brought into focus IN FRONT of the Retina.

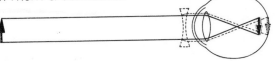

This is shortsightedness or MYOPIA — only objects near the eye can be seen clearly.

It can be corrected by using a concave lens.

RETINA

Sections of the Retina examined under the microscope show 8 layers:-

LAYER of PIGMENT CELLS next to Choroid Coat.

In <u>bright</u> light pigment granules migrate into the cell processes lying between rods and cones. This prevents spread of light from one receptor to another. In <u>dim</u> light the granules are confined to the cell body. When light strikes RODS and CONES, impulses are set up which are transmitted thus:-

RODS and CONES → INTERMEDIATE NEURONES → GANGLION CELLS or INTEGRATING NEURONES → Nerve fibres converge on Optic Papilla to leave as the **OPTIC NERVE**

↓

to Visual Area of **CEREBRAL CORTEX**

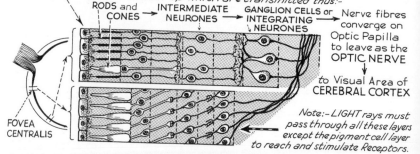

Note:– LIGHT rays must pass through all these layers except the pigment cell layer to reach and stimulate Receptors.

FOVEA CENTRALIS

The <u>FUNDUS OCULI</u> is the part of the Retina seen through the Ophthalmoscope.

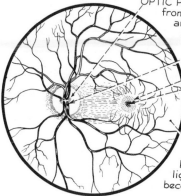

OPTIC PAPILLA – or *'BLIND SPOT'* – The nerve fibres from all parts of the retina converge on this area to leave the eyeball as the OPTIC NERVE. It has no RODS or CONES and therefore is not itself sensitive to light.

RETINAL BLOOD VESSELS enter or leave the eyeball here.

MACULA LUTEA – or *'YELLOW SPOT'* – with FOVEA CENTRALIS – area of acute vision – contains CONES only (the Receptors stimulated in Bright and Coloured light). When we look at an object the eyes are directed so that the image will fall on the fovea of each eye.

EXTRAFOVEAL part of Retina – area of less acute vision – CONES become fewer – RODS (the Receptors stimulated in Dim light with no discrimination between Colours) become more numerous towards the peripheral part of the Retina.

MECHANISM of VISION

WHITE LIGHT is really due to the fusion of *coloured lights*. These coloured lights are separated by shining a beam of white light through a glass prism. This is called the VISIBLE SPECTRUM.

7,800 Å 7,000 Å	6,000 Å			5,000 Å			3,900 Å
RED	ORANGE	YELLOW	YELL.-GR.	GREEN	BLUE-GREEN	BLUE	VIOLET

If light is bright or intense the spectrum appears brightest to man's eye in the orange band (6,100 Å).

If brightness or intensity of light source is gradually reduced, colour perception is gradually lost and the spectrum appears as a luminous band with a very dark area in the red band and with its brightest part in the green band (5,300Å).

Visual Receptors contain PIGMENTS which break down chemically in the presence of specific wavelengths of light. This forms a chemical stimulus which in some way triggers off nerve impulses which travel from the retina to the cerebral cortex.

SCOTOPIC VISION is vision in <u>DIM LIGHT</u>.   It depends on the <u>RODS</u>

<u>RODS</u> are of one type ———————and give————————MONOCHROMATIC VISION

In DIM LIGHT

<u>RODS</u> contain   RHODOPSIN ("VISUAL   *absorbs light*   ⟶ RETINENE + OPSIN
or   PURPLE")   *Pigment gradually splits*   *(a protein)*
SCOTOPSIN

[In BRIGHT LIGHT   *Chemical Stimulus*
RHODOPSIN- very rapid breakdown with   *triggers off*
"bleaching."*]*   *Nerve Impulse*

In DARKNESS
RHODOPSIN ⟵   *regeneration*   RETINENE + OPSIN
                      (regenerated from Vit.A)

When a person passes from a brightly lit scene to darkness he is temporarily blinded. After about ½ hour he sees well. This adjustment or increase in sensitivity is called <u>DARK ADAPTATION</u>.

As light brightness or intensity increases rods lose their sensitivity and cease to respond.

133

MECHANISM of VISION

<u>PHOTOPIC VISION</u> is vision in <u>BRIGHT LIGHT</u>. It depends on the <u>CONES</u>

<u>CONES</u> are thought to be of 3 types —— giving <u>TRICHROMATIC VISION</u>.
Each type with a
different
photosensitive
<u>VISUAL</u>
<u>PIGMENT</u> with its own wavelength to which it is sensitive, which
it absorbs and by which it is broken down to form the
chemical stimulus.

Note:- The
Photosensitive
pigments in
cones have
not yet been
isolated.

"<u>RED</u>"   receptors absorb <u>YELLOW—ORANGE light</u>
"<u>GREEN</u>" receptors absorb   GREEN   light
"<u>BLUE</u>" receptors absorb   BLUE   light

All 3 types of "PHOTOPSINS" are probably stimulated in roughly equal
proportions when <u>WHITE light</u> falls on retina:
2 or more in varying proportions when <u>OTHER COLOURS of light</u> fall on
retina.

The various types of colour blindness could be explained in terms of the
absence or deficiency of one or more of these special receptors.

[A colour sensation has 3 qualities:-
Hue ———— depends largely on wavelength.

Saturation —— purity –
A "saturated" colour has no white light mixed with it.
An "unsaturated" colour has some white light mixed with it.
Intensity —— brightness— depends largely on "strength" of the light.]

As the intensity of light is reduced the Cones cease to respond and
the Rods take over.

When a person passes from darkness to bright light he is dazzled
but after a short time he sees well again.
This adjustment or decrease in sensitivity on exposure to bright light
is called <u>LIGHT ADAPTATION</u>.

VISUAL PATHWAYS to the BRAIN

The RECEPTORS for VISION are linked by a chain of Neurones with RECEIVING and INTEGRATING CENTRES in the OCCIPITAL LOBES of the CEREBRAL CORTEX.

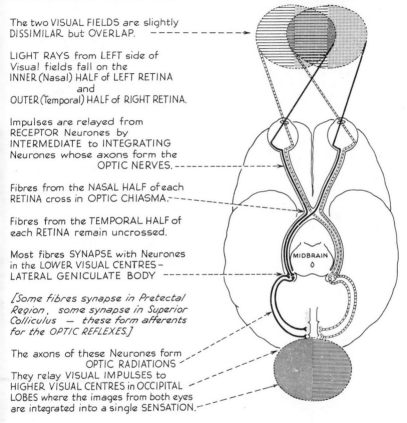

The two VISUAL FIELDS are slightly DISSIMILAR but OVERLAP. ─ ─ ─ ─ ─ ─ ─ ─ ─ ─ ➤

LIGHT RAYS from LEFT side of Visual fields fall on the INNER (Nasal) HALF of LEFT RETINA
                and
OUTER (Temporal) HALF of RIGHT RETINA.

Impulses are relayed from RECEPTOR Neurones by INTERMEDIATE to INTEGRATING Neurones whose axons form the
                OPTIC NERVES. ─ ─ ─ ─ ─ ─ ─ ─ ─ ─ ─ ─ ➤

Fibres from the NASAL HALF of each RETINA cross in OPTIC CHIASMA. ─ ─ ─

Fibres from the TEMPORAL HALF of each RETINA remain uncrossed.

Most fibres SYNAPSE with Neurones in the LOWER VISUAL CENTRES ─ LATERAL GENICULATE BODY ─ ─ ─ ─ ─ ─ ─ ─ ─ ➤

[Some fibres synapse in Pretectal Region, some synapse in Superior Colliculus — these form afferents for the OPTIC REFLEXES.]

The axons of these Neurones form
                OPTIC RADIATIONS
They relay VISUAL IMPULSES to HIGHER VISUAL CENTRES in OCCIPITAL LOBES where the images from both eyes are integrated into a single SENSATION.

Note:- One side of the OCCIPITAL CORTEX receives impressions from the FIELD of VISION on the opposite side.

135

STEREOSCOPIC VISION

When we look at some object or scene the view seen by the RIGHT EYE is slightly different from the view seen by the LEFT EYE.

These two DISSIMILAR RETINAL IMAGES are fused in the Visual Centres of the Brain to give a 3-dimensional picture – an appreciation of DEPTH as well as of HEIGHT and WIDTH.

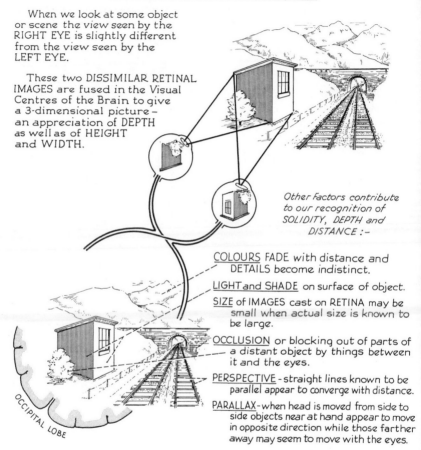

Other Factors contribute to our recognition of SOLIDITY, DEPTH and DISTANCE :–

<u>COLOURS</u> FADE with distance and DETAILS become indistinct.

<u>LIGHT</u> and <u>SHADE</u> on surface of object.

<u>SIZE</u> of IMAGES cast on RETINA may be small when actual size is known to be large.

<u>OCCLUSION</u> or blocking out of parts of a distant object by things between it and the eyes.

<u>PERSPECTIVE</u> - straight lines known to be parallel appear to converge with distance.

<u>PARALLAX</u> - when head is moved from side to side objects near at hand appear to move in opposite direction while those farther away may seem to move with the eyes.

OCCIPITAL LOBE

By complex mental processes these points are interpreted in terms of Distance and Depth.

EAR

The ear has 3 separate parts, each with different rôles in the mechanism of HEARING:-

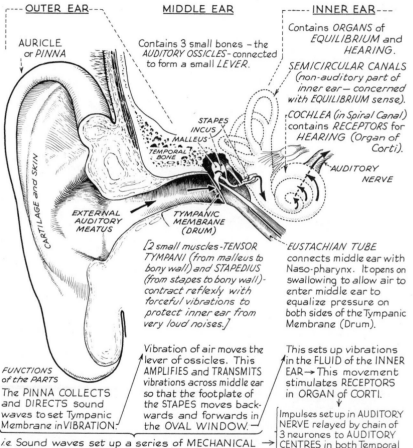

OUTER EAR

AURICLE or PINNA

CARTILAGE and SKIN

EXTERNAL AUDITORY MEATUS

MIDDLE EAR

Contains 3 small bones – the AUDITORY OSSICLES - connected to form a small LEVER.

STAPES
INCUS
MALLEUS
TEMPORAL BONE

TYMPANIC MEMBRANE (DRUM)

[2 small muscles - TENSOR TYMPANI (from malleus to bony wall) and STAPEDIUS (from stapes to bony wall) - contract reflexly with forceful vibrations to protect inner ear from very loud noises.]

INNER EAR

Contains ORGANS of EQUILIBRIUM and HEARING.

SEMICIRCULAR CANALS (non-auditory part of inner ear — concerned with EQUILIBRIUM sense).

COCHLEA (in Spiral Canal) contains RECEPTORS for HEARING (Organ of Corti).

AUDITORY NERVE

EUSTACHIAN TUBE connects middle ear with Naso-pharynx. It opens on swallowing to allow air to enter middle ear to equalize pressure on both sides of the Tympanic Membrane (Drum).

FUNCTIONS of the PARTS

The PINNA COLLECTS and DIRECTS sound waves to set Tympanic Membrane in VIBRATION.

Vibration of air moves the lever of ossicles. This AMPLIFIES and TRANSMITS vibrations across middle ear so that the footplate of the STAPES moves backwards and forwards in the OVAL WINDOW.

This sets up vibrations in the FLUID of the INNER EAR → This movement stimulates RECEPTORS in ORGAN of CORTI.

Impulses set up in AUDITORY NERVE relayed by chain of 3 neurones to AUDITORY CENTRES in both Temporal lobes (see page 117)

i.e. Sound waves set up a series of MECHANICAL → STIMULI

MECHANISM of HEARING

This is most readily understood if the COCHLEA is imagined as straightened out:—

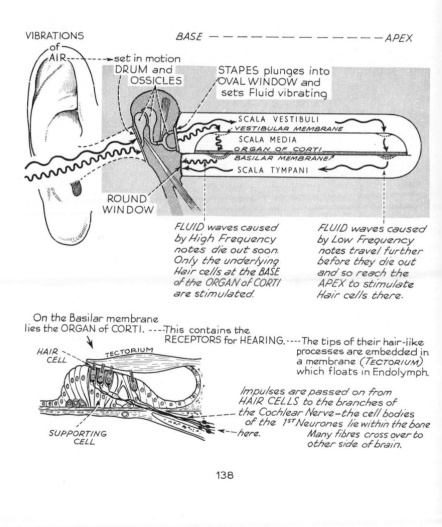

VIBRATIONS of AIR ---- set in motion DRUM and OSSICLES

BASE — — — — — — — — — — APEX

STAPES plunges into OVAL WINDOW and sets Fluid vibrating

SCALA VESTIBULI

VESTIBULAR MEMBRANE

SCALA MEDIA

ORGAN OF CORTI

BASILAR MEMBRANE

SCALA TYMPANI

ROUND WINDOW

FLUID waves caused by High Frequency notes die out soon. Only the underlying Hair cells at the BASE of the ORGAN of CORTI are stimulated.

FLUID waves caused by Low Frequency notes travel further before they die out and so reach the APEX to stimulate Hair cells there.

On the Basilar membrane lies the ORGAN of CORTI. ----This contains the RECEPTORS for HEARING.----The tips of their hair-like processes are embedded in a membrane (TECTORIUM) which floats in Endolymph.

HAIR CELL   TECTORIUM

SUPPORTING CELL

Impulses are passed on from HAIR CELLS to the branches of the Cochlear Nerve – the cell bodies of the 1ST Neurones lie within the bone here. Many fibres cross over to other side of brain.

SPECIAL PROPRIOCEPTORS

The "Special" Proprioceptors of the body are found in the Non-Auditory part of the Inner Ear — the Labyrinth. They are stimulated by movements or change of position of the head in space.

Three **SEMICIRCULAR CANALS** — *in each inner ear — one in each of three planes of space* — contain Receptors.

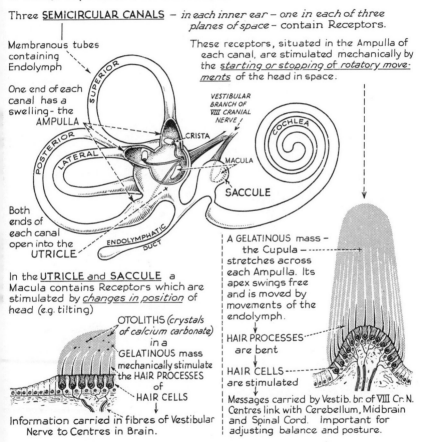

Membranous tubes containing Endolymph

One end of each canal has a swelling - the AMPULLA

SUPERIOR

POSTERIOR

LATERAL

CRISTA

VESTIBULAR BRANCH OF VIII CRANIAL NERVE

COCHLEA

MACULA

SACCULE

Both ends of each canal open into the UTRICLE

ENDOLYMPHATIC DUCT

These receptors, situated in the Ampulla of each canal, are stimulated mechanically by the *starting or stopping of rotatory movements* of the head in space.

In the **UTRICLE** and **SACCULE** a Macula contains Receptors which are stimulated by *changes in position* of head (e.g. tilting)

OTOLITHS *(crystals of calcium carbonate)* in a GELATINOUS mass mechanically stimulate the HAIR PROCESSES of HAIR CELLS

Information carried in fibres of Vestibular Nerve to Centres in Brain.

A GELATINOUS mass — the Cupula — stretches across each Ampulla. Its apex swings free and is moved by movements of the endolymph.

HAIR PROCESSES are bent

HAIR CELLS are stimulated

Messages carried by Vestib. br. of VIII Cr. N. Centres link with Cerebellum, Midbrain and Spinal Cord. Important for adjusting balance and posture.

GENERAL PROPRIOCEPTORS

Proprioceptors are the sense organs stimulated by MOVEMENT of the body itself. They make us aware of the movement or position of the body in space and of the various parts of the body to each other. They are important as ingoing afferent pathways in reflexes for adjusting Posture and Tone. They may be linked with centres in (a) CEREBELLUM or (b) PARIETAL LOBE of opposite side of brain where awareness of muscle and joint sense is appreciated. General Proprioceptors are found in SKELETAL MUSCLES, TENDONS and JOINTS.

GOLGI ORGAN — in tendons —
stimulated by tension which occurs when muscle is STRETCHED and when it is CONTRACTED

Proprioceptor impulses travel from trunk and limbs in spinal nerves

BONE

MUSCLE SPINDLE —
—in skeletal muscle— stimulated when muscle is STRETCHED or SHORTENED

PACINIAN CORPUSCLES
similar to those in the skin are found in Deep Connective Tissue and around Joints. They are stimulated by PRESSURE of surrounding structures when joints are moved.

140

CUTANEOUS SENSATION

There are 5 basic skin sensations – TOUCH, PRESSURE, PAIN, WARMTH and COLD. There is much controversy as to how these are registered. In some areas they appear to be served by special nerve endings (sensory receptors or end-organs) in the SKIN. These end-organs are not uniformly distributed over the whole body surface. (Eg. Touch "endings" are very numerous in hands and feet but are much less frequent in the skin of the back.)

In _HAIRLESS_ parts of the skin (the palms of the hands and soles of the feet)

MEISSNER corpuscles register **TOUCH**.

KRAUSE bulbs (*function uncertain*).

RUFFINI endings (*function uncertain*).

PACINIAN corpuscles deep in dermis are stimulated by **PRESSURE**.

Branching nerve endings in epidermis and dermis register **PAIN** and probably other skin sensations.

In _HAIRY_ surfaces of body

Network of nerve fibres round SHAFT of HAIR registers sensations of **TOUCH** when hair is moved.

Tickling, itching, softness, hardness, wetness are probably due to stimulation of two or more of these special endings and to a blending of the sensations in the Brain.

10

SENSORY PATHWAYS and CORTEX

The nerve endings registering Pain, Warmth or Cold, Touch and Pressure, are linked by a chain of 3 Neurones with final RECEIVING CENTRES in the SENSORY area of the PARIETAL LOBES. The exact points to which impulses come from different regions are indicated.

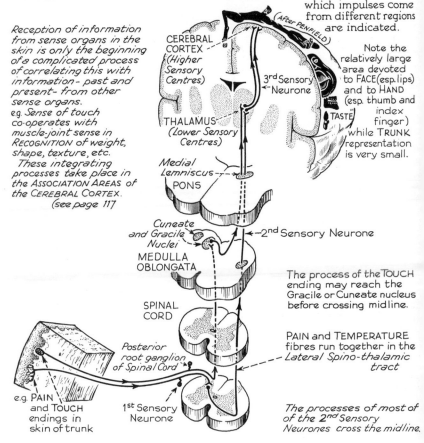

Reception of information from sense organs in the skin is only the beginning of a complicated process of correlating this with information- past and present- from other sense organs.
e.g. Sense of touch co-operates with muscle-joint sense in RECOGNITION of weight, shape, texture, etc.
These integrating processes take place in the ASSOCIATION AREAS of the CEREBRAL CORTEX.
(see page 117

(After PENFIELD)

CEREBRAL CORTEX (Higher Sensory Centres)

3rd Sensory Neurone

Note the relatively large area devoted to FACE (esp. lips) and to HAND (esp. thumb and index finger) while TRUNK representation is very small.

TASTE

THALAMUS (Lower Sensory Centres)

Medial Lemniscus
PONS

Cuneate and Gracile Nuclei
MEDULLA OBLONGATA

2nd Sensory Neurone

SPINAL CORD

The process of the TOUCH ending may reach the Gracile or Cuneate nucleus before crossing midline.

Posterior root ganglion of Spinal Cord

PAIN and TEMPERATURE fibres run together in the Lateral Spino-thalamic tract

e.g. PAIN and TOUCH endings in skin of trunk

1st Sensory Neurone

The processes of most of the 2nd Sensory Neurones cross the midline.

MOTOR PATHWAYS to TRUNK and LIMBS

The CONTROLLING CENTRES in the MOTOR CORTEX are linked by
2 Neurones with the EFFECTOR ORGANS – the VOLUNTARY MUSCLES.

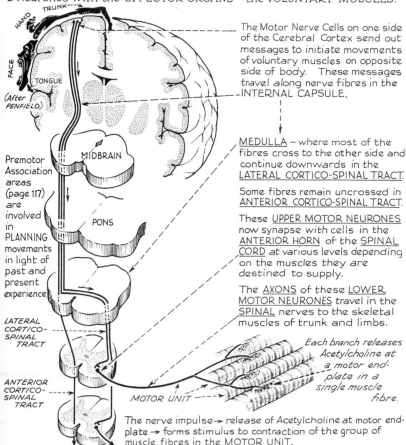

TRUNK
HAND
FACE
TONGUE
(After PENFIELD)

Premotor
Association
areas
(page 117)
are
involved
in
PLANNING
movements
in light of
past and
present
experience

MIDBRAIN

PONS

LATERAL
CORTICO-
SPINAL
TRACT

ANTERIOR
CORTICO-
SPINAL
TRACT

MOTOR UNIT

The Motor Nerve Cells on one side
of the Cerebral Cortex send out
messages to initiate movements
of voluntary muscles on opposite
side of body. These messages
travel along nerve fibres in the
INTERNAL CAPSULE.

MEDULLA – where most of the
fibres cross to the other side and
continue downwards in the
LATERAL CORTICO-SPINAL TRACT.

Some fibres remain uncrossed in
ANTERIOR CORTICO-SPINAL TRACT.

These UPPER MOTOR NEURONES
now synapse with cells in the
ANTERIOR HORN of the SPINAL
CORD at various levels depending
on the muscles they are
destined to supply.

The AXONS of these LOWER
MOTOR NEURONES travel in the
SPINAL nerves to the skeletal
muscles of trunk and limbs.

*Each branch releases
Acetylcholine at
a motor end-
plate in a
single muscle
fibre.*

The nerve impulse→ release of Acetylcholine at motor end-
plate → forms stimulus to contraction of the group of
muscle fibres in the MOTOR UNIT.

CONTROL of MUSCLE MOVEMENT

The Cerebral Cortex initiates purposeful movements. During such movements proprioceptors are continually supplying information to Cerebellum about the changing positions of muscles and joints. The Cerebellum is then responsible for collating this information and sending out impulses to bring about the smooth co-ordinated movements of different groups of muscles.

This involves co-operation of other centres in the

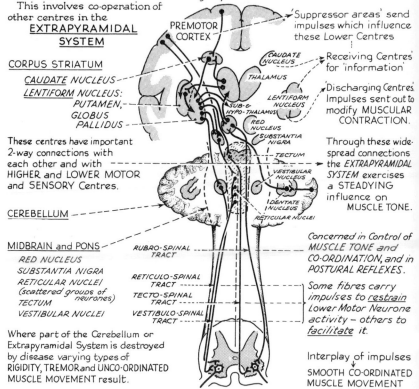

EXTRAPYRAMIDAL SYSTEM

PREMOTOR CORTEX

'Suppressor areas' send impulses which influence these Lower Centres

CORPUS STRIATUM

CAUDATE NUCLEUS
LENTIFORM NUCLEUS:
　　PUTAMEN,
　　GLOBUS
　　PALLIDUS

CAUDATE NUCLEUS

THALAMUS

LENTIFORM NUCLEUS

SUB-& HYPO-THALAMUS

RED NUCLEUS

SUBSTANTIA NIGRA

TECTUM

VESTIBULAR NUCLEUS

DENTATE NUCLEUS

RETICULAR NUCLEI

Receiving Centres for 'information'

Discharging Centres. Impulses sent out to modify MUSCULAR CONTRACTION.

These centres have important 2-way connections with each other and with HIGHER and LOWER MOTOR and SENSORY Centres.

Through these wide-spread connections the *EXTRAPYRAMIDAL SYSTEM* exercises a STEADYING influence on MUSCLE TONE.

CEREBELLUM

MIDBRAIN and PONS

RED NUCLEUS
SUBSTANTIA NIGRA
RETICULAR NUCLEI
(scattered groups of neurones)
TECTUM
VESTIBULAR NUCLEI

RUBRO-SPINAL TRACT

RETICULO-SPINAL TRACT

TECTO-SPINAL TRACT

VESTIBULO-SPINAL TRACT

Concerned in Control of MUSCLE TONE and CO-ORDINATION, and in POSTURAL REFLEXES.

Some fibres carry impulses to restrain Lower Motor Neurone activity – others to facilitate it.

Where part of the Cerebellum or Extrapyramidal System is destroyed by disease varying types of RIGIDITY, TREMOR and UNCO-ORDINATED MUSCLE MOVEMENT result.

Interplay of impulses
↓
SMOOTH CO-ORDINATED MUSCLE MOVEMENT

Little is known about the methods for regulating this interplay of impulses converging on the MOTOR UNIT during a voluntary action.

LOCOMOTOR SYSTEM

BONES and MUSCLES ——— concerned with MOVEMENT of the body.

SKELETON——RIGID FRAMEWORK gives SHAPE and SUPPORT to body.
is JOINTED to permit MOVEMENT

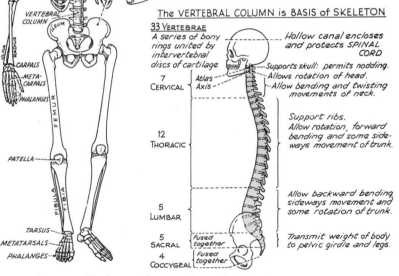

FLAT BONES protect delicate organs.

LONG BONES act as levers.

SHORT BONES confer strength.

The VERTEBRAL COLUMN is BASIS of SKELETON

33 VERTEBRAE

A series of bony rings united by intervertebral discs of cartilage

Hollow canal encloses and protects SPINAL CORD

7 CERVICAL	Atlas Axis	Supports skull: permits nodding. Allows rotation of head. Allow bending and twisting movements of neck.
12 THORACIC		Support ribs. Allow rotation, forward bending and some sideways movement of trunk.
5 LUMBAR		Allow backward bending, sideways movement and some rotation of trunk.
5 SACRAL	Fused together	Transmit weight of body to pelvic girdle and legs.
4 COCCYGEAL	Fused together	

All bones give attachment to muscles.

Labels on skeleton: CLAVICLE, SCAPULA, HUMERUS, RIBS, VERTEBRAL COLUMN, ILEUM, ULNA, RADIUS, CARPALS, META-CARPALS, PHALANGES, FEMUR, PATELLA, FIBULA, TIBIA, TARSUS, METATARSALS, PHALANGES

SKELETAL MUSCLES

Bones are moved at joints by the CONTRACTION and RELAXATION of MUSCLES attached to them.

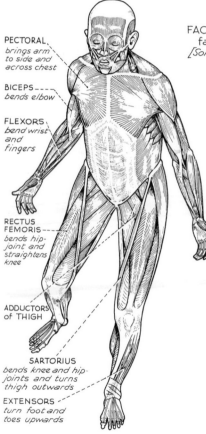

PECTORAL
brings arm to side and across chest

BICEPS
bends elbow

FLEXORS
bend wrist and fingers

RECTUS FEMORIS
bends hip-joint and straightens knee

ADDUCTORS
of THIGH

SARTORIUS
bends knee and hip-joints and turns thigh outwards

EXTENSORS
turn foot and toes upwards

FACIAL muscles are involved in varying facial *Expression; Speech; Mastication [Some muscles link bone to skin.]*

The deep muscles of the THORAX, linking the ribs, contract and relax in *Respiration.*

The muscles of the ABDOMEN are arranged in sheets and *Protect* delicate abdominal organs. They also contract to compress abdominal contents and aid in *Micturition, Defaecation, Vomiting* (and in the processes of *Childbirth* in the female).

In the LEGS are found the most powerful muscles of the body – especially those acting on the hip-joint.

Muscles which bend a limb at a joint are called **FLEXORS**.

Muscles which straighten a limb at a joint are called **EXTENSORS**.

Muscles which move a limb (or other part) away from the midline are called **ABDUCTORS**.

Muscles which move a limb (or other part) towards the midline are called **ADDUCTORS**.

SKELETAL MUSCLES

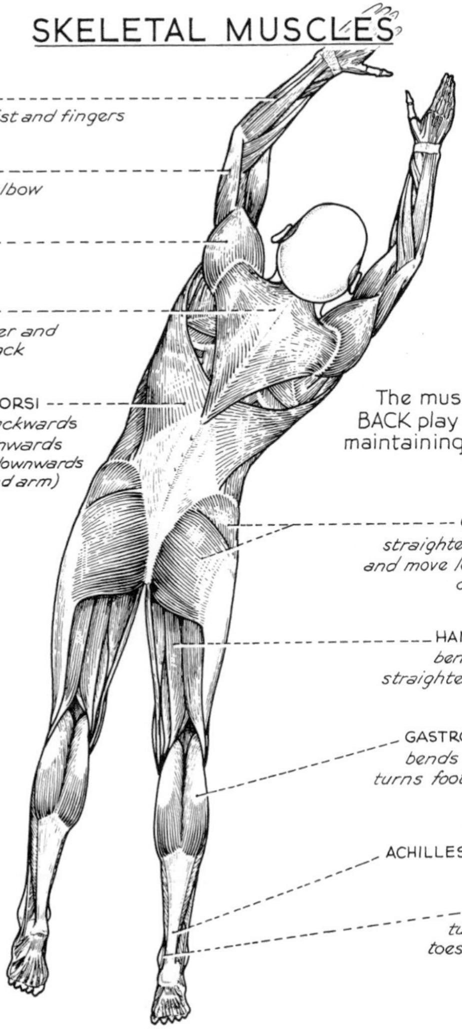

EXTENSORS ----------------
straighten wrist and fingers

TRICEPS ----------------
straightens elbow

DELTOID ----------------
raises arm

TRAPEZIUS ----------------
*raises shoulder and
pulls head back*

LATISSIMUS DORSI ----------
*draws arm backwards
and turns it inwards
(It also draws downwards
an upstretched arm)*

The muscles of the
BACK play a large part in
maintaining erect posture

---------- GLUTEALS
*straighten hip-joint
and move leg
outwards*

---------- HAMSTRINGS
*bend knee and
straighten hip joint*

GASTROCNEMIUS
*bends knee and
turns foot downwards*

ACHILLES TENDON

FLEXORS
*turn foot and
toes downwards*

Some muscles
work together to
ROTATE a limb
or other part of
the body.

147

MUSCULAR MOVEMENTS

The long bones particularly form a light framework of LEVERS.
The skeletal muscles attached to them contract to operate these levers.

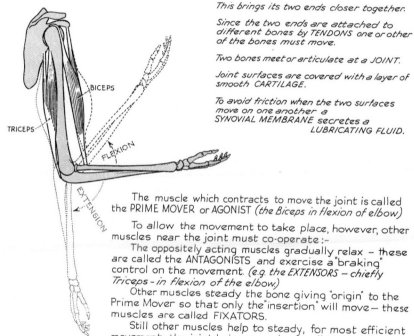

When a muscle contracts it shortens.

This brings its two ends closer together.

Since the two ends are attached to different bones by TENDONS one or other of the bones must move.

Two bones meet or articulate at a JOINT.

Joint surfaces are covered with a layer of smooth CARTILAGE.

To avoid friction when the two surfaces move on one another a SYNOVIAL MEMBRANE secretes a LUBRICATING FLUID.

BICEPS

TRICEPS

FLEXION

EXTENSION

The muscle which contracts to move the joint is called the PRIME MOVER or AGONIST *(the Biceps in flexion of elbow)*

To allow the movement to take place, however, other muscles near the joint must co-operate :-

The oppositely acting muscles gradually relax – these are called the ANTAGONISTS and exercise a "braking" control on the movement. *(e.g. the EXTENSORS – chiefly Triceps - in flexion of the elbow.)*

Other muscles steady the bone giving "origin" to the Prime Mover so that only the "insertion" will move— these muscles are called FIXATORS.

Still other muscles help to steady, for most efficient movement, the joint being moved — called SYNERGISTS.

When the elbow is straightened the reverse occurs:-
TRICEPS, the Prime Mover, CONTRACTS; BICEPS, the Antagonist, RELAXES.

AUTONOMIC NERVOUS SYSTEM

The Autonomic Nervous System is concerned with maintaining a <u>Stable Internal Environment</u>. It governs functions which are normally carried out below the level of consciousness.

It has 2 separate parts – <u>PARASYMPATHETIC</u> and <u>SYMPATHETIC</u>. Both have their Highest Centres in the Brain.

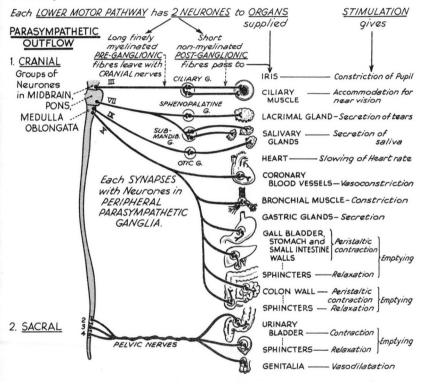

Each <u>LOWER MOTOR PATHWAY</u> has <u>2 NEURONES</u> to <u>ORGANS</u> supplied

<u>STIMULATION</u> gives

PARASYMPATHETIC OUTFLOW

Long finely myelinated <u>PRE-GANGLIONIC</u> fibres leave with CRANIAL nerves

Short non-myelinated <u>POST-GANGLIONIC</u> fibres pass to

1. **CRANIAL** Groups of Neurones in MIDBRAIN, PONS, MEDULLA OBLONGATA

III — CILIARY G.
VII — SPHENOPALATINE G.
IX — SUB-MANDIB. G.
X — OTIC G.

Each SYNAPSES with Neurones in PERIPHERAL PARASYMPATHETIC GANGLIA.

IRIS ———— Constriction of Pupil

CILIARY MUSCLE ———— Accommodation for near vision

LACRIMAL GLAND – Secretion of tears

SALIVARY GLANDS ——— Secretion of saliva

HEART ———— Slowing of Heart rate

CORONARY BLOOD VESSELS – Vasoconstriction

BRONCHIAL MUSCLE – Constriction

GASTRIC GLANDS – Secretion

GALL BLADDER, STOMACH and SMALL INTESTINE WALLS } Peristaltic contraction } Emptying

SPHINCTERS ——— Relaxation

COLON WALL — Peristaltic contraction } Emptying
SPHINCTERS — Relaxation

2. **SACRAL** 2 3 4

PELVIC NERVES

URINARY BLADDER ——— Contraction } Emptying
SPHINCTERS —— Relaxation

GENITALIA ——— Vasodilatation

General stimulation of the Parasympathetic promotes vegetative functions of the body. Stimulation of the Parasympathetic and inhibition of Sympathetic have the same overall effects.

AUTONOMIC NERVOUS SYSTEM

The Parasympathetic and Sympathetic systems normally act in balanced reciprocal fashion. The activity of an organ at any one time is the result of the two opposing influences.

Each <u>LOWER MOTOR PATHWAY</u> has <u>2 NEURONES</u> to <u>ORGANS</u> supplied

STIMULATION gives

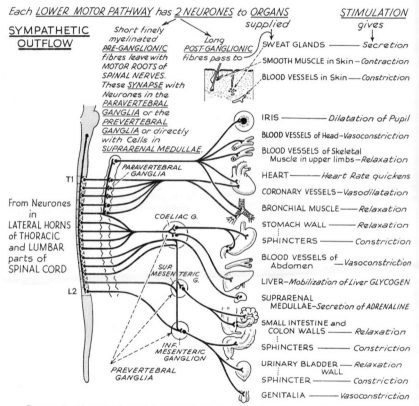

SYMPATHETIC OUTFLOW

Short finely myelinated <u>PRE-GANGLIONIC</u> fibres leave with *MOTOR ROOTS of SPINAL NERVES*. These <u>SYNAPSE</u> with Neurones in the <u>PARAVERTEBRAL GANGLIA</u> or the <u>PREVERTEBRAL GANGLIA</u> or directly with Cells in <u>SUPRARENAL MEDULLAE</u>.

Long <u>POST-GANGLIONIC</u> fibres pass to

SWEAT GLANDS ———— *Secretion*

SMOOTH MUSCLE in Skin – *Contraction*

BLOOD VESSELS in Skin —— *Constriction*

PARAVERTEBRAL GANGLIA

T1

From Neurones in **LATERAL HORNS** of **THORACIC** and **LUMBAR** parts of **SPINAL CORD**

COELIAC G.

SUP. MESENTERIC G.

L2

INF. MESENTERIC GANGLION

PREVERTEBRAL GANGLIA

IRIS ———— *Dilatation of Pupil*

BLOOD VESSELS of Head– *Vasoconstriction*

BLOOD VESSELS of Skeletal Muscle in upper limbs– *Relaxation*

HEART ———— *Heart Rate quickens*

CORONARY VESSELS– *Vasodilatation*

BRONCHIAL MUSCLE ———— *Relaxation*

STOMACH WALL ———— *Relaxation*

SPHINCTERS ———— *Constriction*

BLOOD VESSELS of Abdomen ———— *Vasoconstriction*

LIVER– *Mobilization of Liver GLYCOGEN*

SUPRARENAL MEDULLAE– *Secretion of ADRENALINE*

SMALL INTESTINE and COLON WALLS ———— *Relaxation*

SPHINCTERS ———— *Constriction*

URINARY BLADDER —— *Relaxation* WALL

SPHINCTER ———— *Constriction*

GENITALIA ———— *Vasoconstriction*

General stimulation of Sympathetic results in mobilization of resources to prepare body to meet emergencies. Stimulation of the Sympathetic and inhibition of the Parasympathetic have the same overall effects.

INDEX

Bold figures indicate main reference.

A

Abdomen, muscles of, 146
Abducens nerve (VI), 120, 130
Absorption, 6, 32, 36, 37, **39**, **40**, 43, 62, 90, 91
Accessory nerve (XI), 120
Accommodation for near vision, 129, 131
Acetylcholine, 42, 143
Acromegaly, 97
Adaptation, eye, 133, 134; receptors, 126
Addison's disease, 93
Adipose tissue, 9
Adrenal cortex, 82, 83, **92**, **93**, 95, 97
Adrenaline, Adrenal medulla, (46), 92, **94**, 150
Adrenocorticotrophic hormone (ACTH), 92, **95**, 97
Afterbirth, 111
Agglutination, 63, 64
Air, 69, 73, 74
Albumin, blood, 60
Aldosterone, **82**, **83**, **92**, **93**
Alimentary tract, 27–44
Allergy, 8
Alveoli, Alveolar air, 70, 74, 75
Amino acids, 15, 16, 19, 21, 37, 39, 43, 81
Ammonia, 81
Amniotic cavity, 110
Amoeba, 1, 2
Ampulla, semicircular canals, 139
Ampulla of Vater, 35
Amylase, 35, 37
Anal canal, 27, 40, 41, 42
Androgens, of suprarenal cortex, 92, 93, 97
Antibodies, 8, 60, 65
Antidiuretic hormone (ADH), 82, 83, 98
Antigens, 64, 65
Antihaemophilic globulin, 61
Aorta, 48–50, **53**, 54, **56**
Aortic bodies, sinus, 77
Appendix, 27, **40**
Aqueous humour, 129, 131
Arachnoid, 67
Arterioles, **53**, **54**, 78–80, (149, 150)
Arteries, 53–56
Auditory nerve (VIII), 120, 137, 138
Auerbach's plexus, 30, 31, 37, 39, 40

Auricles of heart, 47–50, 52
Auriculo-ventricular node, 52
Autonomic nervous system, **149**, **150**
Axon, 12, 13, 122

B

Balance, sense of (see Equilibrium)
Baroreceptors, 77
Basal ganglia, 118
Basal metabolism, 24, 25, 87–97, 99
Basophil cells, 8, 95, 97
Beri-beri, 23
Betz cells, 12, 117
Bicarbonate, 68, 75
Bile, 27, 35, 36, 40
Biotin, 23
Bladder (urinary), 78, 85, 86, (149, 150)
Blastocyst, 109
Blind spot, 132
Blood, 8, 47, **60–64**
Blood groups, 63, 64
Blood pressure, 54, 55, 71
Blood vessels, **47**, **53**
Body fluids, 57, 59, 82, 83, 92
Body temperature, **44–46**
Bones, 10, (15), 22, **62**, 88, 90, 91, 95–97
Bowman's capsule, 79, 80
Bowman's glands (nose), 127
Brain (see Cerebral cortex)
Brain stem, 124, 144
Breathing, mechanism of, 72
Broad ligament (uterine), 108
Bronchi, bronchioles, **68–70**, (149, 150)
Brunner's glands, 37
Bulbo-urethral glands, 103
Bundle of His, 52

C

Caecum, 27, 40
Calcium, 15, 21, 32, **61**, (84), 88, **90**, **91**
Calorie, 19, 24, 25, 44
Capillaries, 47, 53, 57
Carbohydrate, 15, 16, 19, **20**, 26, 28, 32, 35, 37, 43, (87–99)

Carbon cycle, 17
Carbon dioxide (see Respiration)
Cardiac cycle, 50
Cardiac muscle, **11**, (47–52)
Cardiac sphincter, 27, 30, 31, **32**, (42)
Carotene, 22
Carotid body, sinus, 77
Cartilage, **10**, 69, 70, (137), 148
Cauda equina, 121
Caudate nucleus, 118, 124, 144
Cell, 1, **3**, 4, 5
Centriole, centrosome, 3, 4
Cerebellum, (12), 118, 119, (139, 140), 144
Cerebral cortex, (12), **116–120**, 124, 127, 135, 136, 142–144
Cerebrospinal fluid, 67
Cervix, uterus, 108–111
Chemical senses, 126–128
Chemoreceptors, 77
Childbirth, 110–112
Chlorophyll, 16
Cholecystokinin, 36
Choleretic action, 36
Cholesterol, 60, 89
Choline, 23
Chordae tendineae, 48
Chorionic villi, 109, 110
Choroid coat of eye, 129
Choroid plexuses, 67, 118
Christmas factor, 61
Chromatin, 3, 4
Chromophils, chromophobes, of anterior pituitary, 95
Chromosomes, 3, 7
Chymotrypsin, 35
Cilia, 6, 69, (70), 108, 109
Ciliary body, ganglion, muscle, nerves, 120, 129, 131, 149
Circulation, 47
Claustrum, 118
Cleavage of zygote, 109
Clotting, 61
Cobalamine, 23
Coccygeal nerves, 121
Cochlea, (120), 137–139
Coeliac ganglion, 150
Cold, 46, (77), 141, (142)
Collagen fibres, 9
Colon, 27, **40**, **41**, 42, (149, 150)
Colour blindness, 134
Cones (retina), 132–134
Conditioned reflexes, 29, 124
Conjunctiva, 129, 130
Connective tissues, 8–10, (23)
Cornea, 129, 131
Corpus albicans, 106
Corpus callosum, 119

Corpus luteum, 106, 107, 113
Corpus striatum, 118, 124, 144
Corticoids, suprarenal, (82, 83), 92, 93, 95
Corticotrophin, 92, 95, 97
Cranial nerves, 120
Creatinine, 60, 81, 84
Cretin, 89
Crypts of Lieberkuhn, 37
Cumulus oophorus, 106
Cuneate nucleus, 142
Cushing's syndrome, 93, 97
Cytoplasm, 1, 3

D

Decidua, 111
Defaecation, 40, 41
Dendrites, 12, 13, 122
Dense fibrous tissue, 9
Deoxycorticosterone, 92
Deoxyribonucleic acid (DNA), 3
Diabetes insipidus, mellitus, 98, 99
Diaphragm, (68), 71, 72, 76, 77
Diastole, 50, 51, 53-56
Diet, 19-26, (62)
Digestion, 27-39
Diiodotyrosine, 88
Diuresis, 82
Duodenum, 27, 35-39, 42, 43, (62)
Dura mater, 67
Dwarfing, 89, 96

E

Ear, 137-139
Elastic fibres, 9, 10, 53, 56, 69, 72
Embryo, 109, 110
Endocardium, 48
Endocrine glands, 87-100
Endolymph, 138, 139
Endolymphatic duct, 139
Endometrium, 108-111, 113
Endoplasmic reticulum, 3
Endothelium, (6), 48, 53
Energy sources, requirements, uses, 20, 24, 25, 44, (88, 89)
Enterogastrone, 33, 34
Enterokinase, 35, 37
Enzymes, 27, 28, 32, 33, 35, 37, 60, 62
Eosinophils, 8, 90, 95
Epididymis, 101, 102
Epithelia, 6, 7

Equilibrium, organs of, 126, 137, 139
Erepsin, 37
Ergastoplasm, 3
Erythrocytes (see Red blood corpuscle)
Eustachian tube, 137
Excretion, (2), 78
Exophthalmos, 89
External auditory meatus, 137
Exteroceptors, 126
Extrapyramidal system, 144
Extrinsic factor, 62
Extrinsic muscles of eye, (120), 130
Extrinsic thromboplastin, 61
Eyes, (120), 129-136, (149, 150)

F

Facial muscles, nerve (VII), 120
Factors V, VII, VIII, IX, X, 61
Faeces, 36, 40, 41, 44, 59
Fallopian tubes (uterine tubes), (6), 108, 109, 114
Fat, 15, 16, 19, 20, 26, 32, 35, 36, 37, 39, 43, 45, 46, (99)
Fatty acids, 19, 35, 36, 37, 39, 43
Female (see Reproductive system)
Fertilization, 109, 113
Fibrin(-ogen), 60, 61
Fibrous tissue, 9
' Fight or Flight ' mechanism, 94
Filtration (see Kidney)
Fixators, 148
Foetus, (64), 110
Folic acid, 23, 62
Follicle-stimulating hormone (FSH), 95, 113
Food, 19-26, (27-43)
Fovea centralis, 132
Frohlich's dwarf, 96
Fundus oculi, 132
Fundus of stomach, 32, 34

G

Gall bladder, 27, 36, (149)
Gastric (glands) juice, 32, 33
Gastrin, 33
Gastro-colic reflex, 41
Gastro-intestinal hormones, 33-37, 39, (87)
Gastro-intestinal tract, 27-42
Gastrozymin, 33

Genes, 3, 4
Geniculate bodies, 135
Giantism, 97
Glands (see under specific sites)
Globulin, 60, 65
Globus pallidus, 118, 144
Glomerulus, kidney, 78-81
Glossopharyngeal nerve (IX), 120
Glottis, 69, 77
Glucagon, 99
Glucocorticoids, 92, 93
Glucose, 20, 32, 35, 37, 39, 43, 81, 92, 93, 99
Glycerides, 35-37, 39
Glycerol, 37, 39, 43
Glycogen, 20, 43, 92, 94, 99, (108)
Goblet cells, 6, 40, 69
Golgi body, 3
Golgi organ (in tendons), 140
Gonadotrophins, 95, 104, 113-115
Graafian follicle, 106, 113, 114
Gracile nucleus, 142
Granular beucocytes, 8, 62
Grey matter, 117, 118, 121
Growth, (19), 21-24, 25, 43, 87-100, 104, 114
Growth hormone, 95-97
Gustatory cells, 128

H

Haemoglobin, 21, 62, (63), 75
Haemolysis, 63, 64
Haemolytic disease of newborn, 64
Haemopoiesis, 62, 65, 66
Haemopoietic factor, 62
Haemostasis, 61
Hageman factor, 61
Hair, 45, 46, 89, 93, 94, 97, 101, 104, 105, 114, 115, 141
Hassall's corpuscles, 100
Haversian canal (in bone), 10
Hearing, 117, 120, 137, 138
Heart, 11, 47-58, (89, 94, 97, 149, 150)
Heat, 44-46, (89), 141
Henle's loop (kidney), 79, 81
Heparin, 8
Hering-Breuer reflex, 77
Hirsutism, 93, 97
Hormones, 33-37, 39, 87-115
Hyaloplasm, 3
Hydrochloric acid, in gastric juice, 32, 33
Hydrocortisone, 92, 93

152

Hypermetropia, 131
Hypoglossal nerve (XII), 120
Hypothalamus, 45, 46, 82, 83, 88, 92, 94, 95, 98, 111, 112, 119, 144

I

Ileo-caecal valve, 27, 37, 38, 40, 41, 42
Ileum, 27, 37
Immunity, 8, (60), 65, **100**
Implantation, 109, 110
Incisura angularis, 34
Incus, centres for smell, 127
Incus, middle ear, 137
Inferior mesenteric ganglion, 150
Inositol, 23
Inspiration, 72-77
Insula, 118
Insulin, 99
Intercalated discs, 11
Intercostal muscles, 71, 72
Internal capsule, 118, 142, 143
Interoceptors, 126
Interstitial cells, 102, 104, (114)
Interstitial cell stimulating hormone (ICSH), 104
Intervertebral discs, 10, 145
Intestinal gastrin, 33
Intestine, (6), 27, 34-43, 90, 91, (149), 150
Intrapleural intrathoracic, pressure, (58), 71, 72
Intrinsic factor, 32, 62
Intrinsic thromboplastin, 61
Involuntary muscle, 11
Iodine, 21, 26, 88
Iris, 94, 129, (149, 150)
Iron, 15, 21, 26, 32, (43), 60, **62**
Islets of Langerhans, 35, **99**

J

Jejunum, 27, 37, 38, 39, 42, 43
Joints, 140, 145, **148**

K

Ketone bodies, 99
Kidney, **78-83**, 90-93, (97), 98, 99
Krause bulbs, in skin, 141
Kupffer cells (liver), 36

L

Labyrinth, 139
Lacrimal ducts, glands, sac, 130
Lactase, lactose, 37
Lactation, 25, 95, **112**
Lacteals, 39, 43
Lactogenic hormone, 95
Large intestine, 27, **40-43**, (149, 150)
Larynx, (31), 68, **69**, 120
Lemnisci, 142
Lens of eye, 129, 131
Leucocytes, leucopoiesis, **8**, 60, 62, 65, 66
Leydig, cells of, 102
Ligaments, histology of, 9
Lipase, 32, 35-37
Liquor folliculi, 106
Liver, (20), (27), **36**, 43, (44), 62, 92-94, 99, 150
Locomotor system, 14, 145-148
Loop of Henle, 79, 81
Loose fibrous tissue, 9
Lorain dwarf, 96
Lumbar nerves, 121
Lungs, 44, 47, 59, **68-77**, (149, 150)
Luteinizing hormone (LH), 95, 104, 113-115
Luteotrophin (luteotrophic hormone), 95, 112-115
Lymph, nodes, 8, 43, **65**
Lymphatics, 39, 43, 65
Lymphocytes, lymphoid tissue, 8, 37, 40, 65, 66, 100
Lysosomes, 3

M

Macrophage cells, (8), 65, (66)
Macula lutea, 132
Male, (92), 101-104
Malleus, 137, (138)
Maltase, maltose, (28), 35, 37
Mammary glands, (95, 98, 105), 112, (114), 115
Mastication, 28
Medulla oblogata, 29, 31, 33, 42, 76, 77, **118-121**, **142**, **143**, 144, **149**, (150)
Meiosis, 4, 102, 106
Meissner corpuscles, in skin, 13, 141
Meissner's nerve plexus, in gut, 39
Membrana granulosa, 106
Meningeal membrane, 67
Menopause, 115

Menstrual cycle, 106, **108**, **113**, 114, 115
Mesentery, of gut, 39, 40
Metabolism, 23, 24, 25, 43, 44, (59), 81, 87-99
Metaphase, 4
Micturition, 86
Midbrain, 118-120, 143, 144, 149
Milk, 20-23, 26, 32, 35, (112)
Mineralocorticoids, 92, 93, 97
Minerals, 15, 19, 21, 26, 32, 39, (40), 43, 57, 59, (81), 82, 83, 88, 90-93
Mitochondria, 3
Mitosis, 4, (102)
Mitral valve, 48, 51
Monochromatic vision, 133
Monocyte, 8
Monoiodotyrosine, 88
Motor end-plates, 12, 143
Motor nerve cells, 12, 117, (121), (143)
Mouth, 28
Mucous glands, 28, 30, 69
Multipolar neurones, 12
Muscles, 11, 12, 45, 46, 94, 140, 143, 144, 146-148 (see under Smooth, Skeletal and Cardiac muscle)
Muscle spindle, (123), 140
Muscularis mucosae, 39
Myelin, 12
Myeloblast, 62
Myocardium, 48
Myometrium, **108-111**, (114, 115)
Myopia, 131
Myxoedema, 89

N

Nasal passages, sinuses, 68, 127
Naso-lacrimal duct, 130
Nephrons, 79-81
Nerves, nervous tissues, 12, 13, 122
Neurilemma, 12
Neuroglia, 12
Neuromuscular excitability, 91
Neuromuscular junction, (12), 143
Neurones, **12, 13**, 117, **122**
Neutrophil, 8, 62
Nicotinic acid, 23
Nitrogen, 15, 18, 19, 95, 97
Nodal tissue, heart, 52
Node of Ranvier, 12, 13
Non-granular leucocytes, **8**, 65, 66
Noradrenaline, 42, 94

153

Normoblast, 62
Nose, 68, 120, 126, **127**
Nucleic acids, 62
Nucleus, nucleolus, nucleoplasm, 1, 3, 4
Nutrition, 19-26, (62)

O

Obesity, 93, 96, 97
Oculomotor nerve (III), 120, (130), 149
Oddi, sphincter of, 27, **35, 36,** 149
Oedema, 93
Oesophagus, 10, 27, **30, 31,** 42
Oestrogen, 92, (93), 95, 104, 105, **106,** 109, 110, 112-115
Olfactory bulb, nerve and receptors, 120, (126), **127**
Oocyte, oogenesis, 106
Optic chiasma, nerve, papilla, 120, 129, 130, 132, 135
Organ of Corti, 137, **138**
Organelles, 3
Osmoreceptors, 82, 83, 98
Osmotic pressure, 57, 60, 71, 80, 82, 83, (92), 93, 98
Osteitis fibrosa cystica, 91
Osteomalacia, 22
Otoliths, 139
Ovary, ovarian hormones, (92), 95, 105-108, 112-115
Ovum, ovulation, (4), 106, 108, 109, 113, 114
Oxidation, 19, 43-46, 88, 89
Oxygen, 8, 15, (60), 68, **74,** 75
Oxyntic cells, 32
Oxytocin, 98, 111, 112

P

Pacemaker (S. A. node), 52
Pacinian corpuscles, 13, 140, 141
Pain, (77), 118, 126, 142
Palate, 27, 31, 120
Pancreas, 27, **35, 99**
Pancreozymin, 35
Paneth cells, 37
Pantothenic acid, 23
Papillae of tongue, 128
Papillary muscles, 48
Para-aminobenzoic acid, 23
Parafollicular cells, 88
Parallax, 136
Parasympathetic nervous system (31), (33), 42, 52, 86, **149**
Parathyroid, 90, 91

Paravertebral ganglia, 150
Parotid salivary gland, 28, (120)
Parturition, 110, 111
Pellagra, 23
Pelvic nerves, 41, 42, 86, 150
Penis, 101, **103,** 104
Pepsin, peptic cells, 32
Peptidases, 35, 37
Pericardium, 48
Peripheral resistance, 54
Peristalsis, 30, 31, 34, 38, 40, 41, 108, 109
Pernicious anaemia, 23
Peyer's patch, 37
Phagocytes, 8, 65, 66
Pharynx, 27, 30, 31, 120
Phosphorous, phosphate, 15, 22, 26, (39), (43), 81, 90, 91
Photopic vision, photopsin, 134
Photosynthesis, 16, 17
Pia mater, 67
Pituitary, 82, 83, 87-89, 92, **95-99,** 102, 104, **111-115**
Placenta, 109-113
Plasma proteins, 57, 60, 61, 80
Plasma thromboplastin antecedent, 62
Platelets, 8, 60, 61
Pleura, 70, 71, 72
Pneumotaxic centres, 76
Pons, 76, **118, 119,** 142, 143, (144), 149
Portal canal, 36
Portal vein, 43
Posterior root ganglion, 13, 121, 123, 142
Posture, (139, 140), 144
Pregnancy, 107, 109-113
Prevertebral ganglia, 150
Progesterone, (92), (95), **105-107,** 110-115
Prolactin, 95, 112
Pronormoblast, 62
Prophase, 4
Proprioceptors, 126, 139, 140, (144)
Prostrate gland, 101, 103, 104
Protein, 15, 16, 19, **21,** 26, 32, 35, 43, 45, 46, 62
Prothrombin, 22, 61
Protoplasm, 1, 15
Pseudopodia, 1, 2
Ptyalin, 28
Puberty, 24, 25, 100, 104, 114
Pudendal nerves, 41, 42, 86
Puerperium, 111
Pulmonary circulation, 47, 48, 49, 70, 71, 75
Pulse (radial), 56
Pupil of eye, 129, (130), 149, 150
Purkinje cell, 12
Purkinje tissue, 52
Putamen, 118, 144

Pyloric glands, sphincter, 32, 34, (149, 150)
Pyramidal tract, 119
Pyridoxine, 23

R

Ranvier, node of, 12
Reabsorption, kidney, 81-83, 90-93
Receptors, (13, 38, 77, 82, 83, 86), **126-142**
Rectum, 27, **40-42,** 149, 150
Red blood corpuscles, 8, 60, 62-64, (66), 68, 75
Red nucleus, 144
Reflex action, 29, 31, 33, 35, 36, 38, 41, 77, 86, 112, **123-126,** (131)
Renal corpuscles, tubules, 79-81
Rennin, 32, 35
Reproduction, (2), (4), 92, 93, 95, 98, **101-115**
Respiration, (2), (7), 17, 31, 44, 58, 59, 65, **63-77,** 88, 89, 94, (149, 150)
Rete testis, 102
Reticular nuclei (of brain stem), tract, 144
Reticular tissue, 8, 65, 66
Reticulocyte, 62
Retina, (120), 129, 131, **132-136**
Retinene, 133
Rhesus factor, 64
Rhodopsin, 133
Riboflavine, 23
Ribonucleic acid (RNA), 3
Ribosomes, 3
Ribs, (68), **71, 72,** 145 '
Rickets, 22
Rods (retina), 132
Rubro-spinal tract, 144
Ruffini endings, in skin, 141

S

Saccule, 139
Sacral nerves, (42, 86), 121, 149
Saliva, 27, 28, 29, 32, 59, 83, 120, 128, 149
Sarcolemma, 11
Scala, media, tympani, vestibuli, 138
Schwann cell, 12
Sclerotic coat of eye, 129, 130

154

Scotopic vision, scotopsin, 133
Scrotum, 101-104
Scurvy, 23
Secretin, 35, 37
Segmentation, in small intestine, 38
Semicircular canals, (120), 137, 139
Semilunar valves, 48-50
Seminal fluid, vesicles, 101, 103
Seminiferous tubules, 102, 104
Sense organs, 126
Serous acini, 28, 35
Sertoli, cells of, 102
Sex organs and hormones, 92, 93, 95, **101-115**
Simmond's disease, 96
Sino-auricular node, 52
Skeletal muscle, 11, (12), 30, (40-42), 44-46, (58), (65), 71, 72, (85, 86), 92-94, 97, 99, 123, 125, 130, 140, 143, 144, 146-150
Skeleton, 145
Skin, 7, 22, 44-46, 59, 89, 93, 94, 97, 120, 126, **141**, 142
Small intestine, 27, 37-39, 42, 43
Smell, 118, 120, 126, **127**
Smooth muscle (visceral), **11**, 30-32, 37, 40, 53, 66, 69, 85, 86, 94, 98, 103, 108-111, 114, 115, 149, 150
Sodium, 15, 26, 60, **81-84, 92, 93**
Somatotrophin, **95-97**, (99)
Specific dynamic action of food, 24
Speech, (28), **69**, 77, **117**
Spermatogenesis, spermatozoa, (4), 101, **102**, 104, (109)
Sphincters, 27, 30-32, 34-42, 85, 86, 149, 150
Sphygmomanometer, 55
Spinal cord, (12, 13), 67, (76), (116), **120-125**, **142, 143, 144, 149, 150**
Spinal nerves, (13), **121**, (149, 150)
Spino-thalamic tracts, 142
Spleen, (8), 66
Spino-thalamic tracts, 142
Stapedius muscle, **137**, 138
Starch, 15, (16), (19), **20**, 26, 28, 32, 35, 37, 43
Stercobilin(-ogen), 36
Stereoscopic vision, 136
Sternum, 71, 72
Steroids, 92
Stomach, 27, **32, 33, 34**, 42, 43, 62, 149, 150
Stratified squamous epithelium, **7**, 30, 108, 141

Stress, 77, 92, 94, 95, (112)
Stretch receptors, (38), 77, 123, **140**
Striated border epithelium, 6, 37, 39, 40, 79, 80
Striated muscle (see Skeletal muscle)
Stuart-Prower factor, 61
Subarachnoid space, 67
Sublingual, submandibular, salivary glands (and ganglion), 28, (120, 149)
Substantia nigra, 144
Succus, entericus, 37
Suckling, 111, 112
Sucrase, intestinal, 37
Sugars, **15**, 16, **20**, 26, 32, 35, 37, **39, 43**, 92-94, 99
Supra-optic nucleus, 82, 83, **98**
Suprarenal, (46), 82, 83, **92-95**, 97, 99, (150)
Swallowing, (28), 30, **31**, 77
Sweat (glands), 44-46, 59, 83, 150
Sympathetic ganglion cell, 12
Sympathetic nervous system, 41, 42, 45, 46, 86, 94, 150
Synapse, 122, 123
Synergists, 148
Synovial membrane, 148
Systole, 50-56

T

Tarsal glands, 130
Taste, 28, 29, 120, (126), **128**, 142
Tears, 130
Tectorium of cochlea, 138
Tecto-spinal tract, 144
Teeth, 27, 28
Telereceptors, 126
Telophase, 4
Temperature, 44-46, 141, 142
Tendon, 9, 140
Tensor tympani, 137, (138)
Testis, testosterone, 87, 92, (93), 95, **101-104**
Tetany, 91
Tetraiodothyronine, 88
Thalamus, **118**, 124, 142-144
Thiamine, 23
Thirst, **83**, 99
Thoracic cord and nerves, (42), 76, 121, 150
Thoracic duct, 43, 65
Thorax, 68, **71, 72**, 145

Thrombin, thromboplastin, 61
Thymus, thymocytes, 103
Thyrocalcitonin, 88
Thyroid, thyroglobulin, thyroxine, 88, 89
Thyrotrophin, 88, 95
Tissue fluids, 44, 57, 59
Tissues, 5-15
Tone, 34, 41, 45, 46, 54, 58, 140, 144
Tongue, 27, 28, 29, 31, 120, 128
Touch, (13), 141, 142
Trachea, 68, 69, (88, 90)
Transitional epithelium, 7, 85
Trichromatic vision, 134
Tricuspid valve, 48, 51
Trigeminal nerve (V), 120
Triiodothyronine, 88
Trochlear nerve (IV), 120, 130
Trypsin(-ogen), 35, 37
Tympanic membrane, 137, 138
Tyrosine, 88

U

Umbilical blood vessels and cord, 110, 111
Unstriped muscle (see Smooth muscle)
Urea, uric acid, 60
Ureter, urethra, urinary bladder, 78, **85**, 86, 149, 150
Urinary tract, urine, (44, 59), **78-86**, (91-94, 97-99)
Urobilin(-ogen), 36
Uterus, uterine tubes, 105, 108-111, 113-115
Utricle, 139

V

Vagina, 105, **108**, 114, 115
Vagus nerve (X), 31, 33-36, 42, (52), 76, 77, **120**, 149
Valves, (37, 40), **47-51**, 53, 56, 58, 65
Vas deferens, 101-104
Veins, 43, 47, **53, 54, 58**, 70, 79, 98

155

Ventricles of brain, (67), 118, 119
Ventricles of heart, 47-50, 52
Venules, 53, 54, 58
Vermiform appendix, 40
Vertebrae, 68, 72, 145
Vestibular membrane, nerve, tract, 120, 138, 139, 144
Villi, chorionic, 109, 110, intestinal, 37, 39
Villikinin, 39
Virilism, 93, 97
Vision, 120, **129-136**
Visual purple, 22, **133**
Vital capacity, 73

Vitamins, 22, 23, 43, 62, (81), (133)
Vitreous humour, 129
Vocal cords, 69
Voluntary muscle (see Skeletal muscle)
Von Ebner's glands, 128

W

Warmth (see Temperature)
Water, 15, 26, 32, 39, 40, 43, 57, 59, 80-83, (92, 93), 98, (99)

White blood corpuscles, **8**, 60, 62, 65, 66
Work, 24, 25, 44

X

Xerophthalmia, 22

Z

Zygote, 109

PRINTED IN GREAT BRITAIN
BY GILMOUR & DEAN LTD. HAMILTON AND LONDON